Introduction to high performance liquid chromatography

Introduction to high performance liquid chromatography

R. J. Hamilton
and
P. A. Sewell

Liverpool Polytechnic

SECOND EDITION

London   New York
CHAPMAN AND HALL

First published 1977
by Chapman and Hall Ltd
11 New Fetter Lane, London EC4P 4EE
Second edition 1982
Published in the USA by
Chapman and Hall
in association with Methuen Inc.
733 Third Avenue, New York, NY 10017

Printed in Great Britain by J. W. Arrowsmith Ltd., Bristol

ISBN 0 412 23430 0

British Library Cataloging in Publication Data

Hamilton, R.J. (Richard John)
Introduction to high performance liquid
chromatography.
Includes index.
1. Liquid chromatography.   I. Sewell, P.A.
(Peter Alexis)   II. Title.
QD79.C454H35   1981   534'.0894   81–16840
ISBN 0–412–23430–0   AACR2

Library of Congress Cataloging in Publication Data

Hamilton, R. J. (Richard John)
Introduction to high performance liquid
chromatography.
Includes index.
1. Liquid chromatography.   I. Sewell, P. A.
(Peter Alexis)   II. Title.
QD79.C454H35 1981   543'.0894   81–16840
ISBN 0–412–23430–0   AACR2

Contents

to our wives Shiela and Pat

Preface

Since the first edition of this book the major advances have been in column packings, where over ninety per cent of separations are now performed using chemically bonded microparticulate packings, and in instrumentation.

The use of microprocessor control has brought about a rationalization of mobile phase delivery systems and in detectors, the introduction of electrochemical and spectrophotometric detection other than in the ultra-violet region, has widened the field of applications and the sensitivity of the technique. The use of ion-pair chromatography has increased at the expense of ion-exchange and this together with the improvements in detectors has greatly increased the application of the technique in the biomedical field.

These advances are described together with the established methods to enable the beginner to carry out a satisfactory separation and to gain the experience necessary for the full exploitation of the technique.

R.J. Hamilton
P.A. Sewell

Liverpool, 1981

1

Introduction to high performance liquid chromatography

1.1 Introduction

Chromatography in its many forms is widely used as a separative and an analytical technique. Gas chromatography since its introduction by James and Martin [1] has been pre-eminent in the field. Liquid chromatography in the form of paper, thin-layer, ion-exchange, and exclusion (gel permeation and gel filtration) chromatography had not been able to achieve the same success, mainly because of the poor efficiences and the long analysis times arising from the low mobile phase flow rates. The emergence of liquid chromatography on a basis comparable to gas chromatography is usually considered to start with the publication by Huber and Hulsman [2] in 1967, although Giddings [3] had already shown the potential advantage, in terms of column efficiencies and speed of analysis, of liquid chromatography over gas chromatography.

Various names have been used to describe the main attributes of this 'new' liquid chromatography: high speed (HSLC), high efficiency (HELC), and high pressure or high performance (HPLC). The generally accepted name is now *high performance liquid chromatography* (*HPLC*). It should be made clear that these names refer to the analogue of gas chromatography where the stationary phase, be it a solid surface, a liquid, an ion exchange resin, or a porous polymer, is held in a metal column and the liquid mobile phase is forced through under pressure. Open bed chromatography (paper and thin-layer chromatography) is not included although claims are being made for so-called high performance thin-layer chromatography (HPTLC). These forms of chromatography, together with the 'classical' low pressure column chromatography, will continue as part of the analysts armoury and their utility should not be overlooked. Affinity chromatography, using specific biological interactions, is used increasingly for the separation of high-molecular-weight biological substances, but this too is beyond the scope of this book.

Both gas and high performance liquid chromatography have their place in the analytical laboratory, and there will obviously be an area of overlap where either technique could be used. In general, however, capital costs for liquid chromatography equipment, and the running costs of column packings and mobile phases, are far higher than for gas chromatography, so gas chromatography will probably remain the preferred technique in these areas of overlap.

However, a large number of organic compounds are too unstable or are insufficiently volatile to be handled by GC without prior chemical modification, and liquid chromatography would be the first choice for such compounds. It is ideally suited for the separation of a wide range of pharmaceuticals, food, heavy industrial, and bio-chemicals.

Because lower temperatures can be used, and because there are two competing phases (mobile and stationary) compared with one phase (the stationary phase) in GC, liquid chromatography may often achieve separations that are impossible by GC. Furthermore, there is a wide choice of detectors available for use in LC, many of which are selective, so a complete separation need not necessarily be made on the column but a detector can be chosen that will monitor only species of interest.

Finally, recovery of the sample in LC can be achieved more easily and quantitatively than in GC. Although the mobile phase in LC has to be removed by distillation or some other means, this does not usually present any difficulty because of the wide difference in volatility between the mobile phase and the sample.

The advantages of HPLC over other forms of liquid chromatography may be summarized thus: (i) the HPLC column can be used many times without regeneration; (ii) the resolution achieved on such columns far exceeds that of the older methods; (iii) the technique is less dependent on the operator's skill, and reproducibility is greatly improved; (iv) the instrumentation of HPLC lends itself to automation and quantitation; (v) analysis times are generally much shorter; (vi) preparative liquid chromatography is possible on a much larger scale.

1.2 Nomenclature

Because of its relationship both to older 'classical' forms of liquid chromatography (column, thin-layer, and paper) and to gas chromatography, some confusion may exist in the language of high performance liquid chromatography.

In classical forms of liquid—solid chromatography the *sample* (or *solute*) was dissolved in a *solvent* and was *eluted* from a packed column containing silica gel or alumina. In gas—liquid chromatography the sample is carried through the column by the *carrier gas* (or *mobile phase*) and retention occurs on the *stationary phase*. Because the thermodynamics of the separation involves a simple two-component process, sample and stationary phase are often equated with the terms *solute* and *solvent*. Hence the term 'solvent' has two different meanings in the context of the two techniques.

In this book we shall use the term *solute* or *sample* to represent the components of the mixture to be separated, *stationary phase* or more specifically *adsorbent* or *absorbent* to represent the column packing on which the separation takes place, and *solvent, mobile phase*, or *eluent* to represent the eluting agent.

1.3   Liquid Chromatography Modes

One of the major advantages of liquid chromatography over other separation techniques is to be found in the several different mechanisms by which the chromatographic separation may be achieved. These mechanisms or modes of operation make it possible to achieve separations by liquid chromatography within such diverse sample types as solvents, ionic compounds and polymers. Although the modes of chromatography shall be discussed individually in practice one or more modes may be responsible for effecting the separation. This of course may add another dimension to a separation, but it also makes the prediction of retention behaviour more speculative.

1.3.1   Liquid–liquid (Partition) Chromatography (LLC)

Liquid–liquid or partition chromatography was developed by Martin and Synge [4] in 1941 for the separation of acetylated amino acids using a stationary phase of water on silica gel with chloroform as mobile phase. LLC involves the use of a liquid stationary phase, which is either coated on to a finely divided inert support or chemically bonded to the support material, and a liquid mobile phase. The sample to be analysed is dispersed in the mobile phase and its components are partitioned between the stationary and mobile phases according to their partition coefficients $K_1, K_2 \cdots K_n$. This partitioning leads to a differential rate of migration and separation occurs.

The early workers in high performance liquid–liquid chromatography had their origins in 'classical' and in gas chromatography where the coating procedure for preparing the stationary phase is standard practice. The use of stationary phases prepared in this way in high performance liquid chromatography is unsatisfactory because the liquid stationary phase is stripped off the column, either by solubility effects or by mechanical shear forces, resulting in changes in retention times on the column. To prevent 'stripping', due to solubility effects the mobile phase could be passed through a pre-column containing the same stationary phase as the analytical column, so that it becomes saturated with respect to the stationary phase, and no further stationary phase is removed from the analytical column. However, the presence of a pre-column excludes the use of the gradient elution technique, since the phases would not be in equilibrum, an essential requirement in chromatography.

The development of chemically bonded stationary phases or 'brushes' by Halasz and Sebastian [5] eliminates the problem of stripping of the stationary phase and allows the use of gradient elution in the normal way.

The original bonded phases were prepared by reacting the surface —OH groups of silica with an alcohol to give an Si—O—R grouping. Due to the easy hydrolysis of this grouping these stationary phases could only be used in the pH range 4–7.

The majority of modern bonded phases are prepared from silica by reacting the surface silanol groups with an organochlorosilane or alkoxysilane to give an

$$\begin{array}{c} | \\ Si-O-Si-R \\ | \end{array}$$

linkage which is hydrolytically stable. The R group may be a hydrocarbon (e.g. C_8 or C_{18}) or a hydrocarbon with a polar terminal group (e.g. $R-CN$, $R-NH_2$). The ability to prepare bonded phases with functional groups of varying polarity now means that bonded phases may range from polar to non-polar. Furthermore, by introducing a polar group, such as nitrile, a stationary phase with many of the characteristics, but none of the disadvantages, of silica may be produced.

The success of bonded phases has virtually eliminated the use of coated stationary phases.

The term *normal phase* liquid–liquid chromatography originally referred to a system with a polar liquid stationary phase, e.g. water, glycol or β,β' -oxydipropionitrile, while the mobile phase is relatively non-polar, e.g. hexane, benzene, or chloroform. This mode of operation was used to separate polar compounds which would be distributed preferentially in the polar stationary phase.

If the stationary phase is non-polar, e.g. a hydrocarbon, and the mobile phase is polar, e.g. water, the technique is referred to as *reverse-phase* liquid chromatography. In reverse-phase chromatography the non-polar stationary phase may be displaced from a polar support by virtue of the preferential adsorption of the polar mobile phase. This may be prevented by de-activating the support material by silanization before coating with the stationary phase.

With the advent of bonded phases the distinction between normal and reversed-phase chromatography has been redefined. Normal phase refers to a system where the stationary phase is more polar than the mobile phase and reversed phase where the stationary phase is less polar than the mobile phase. Since the majority of separations are now carried out in the reversed-phase mode, the choice of name is a little unfortunate.

Ion pair chromatography (IPC) is a special form of LLC used for the separation of ionic or ionizable compounds, e.g. quaternary ammonium salts, sulphonates, amino acids, and amino-phenols [6]. In its most usual form it is used in the reverse-phase mode with a hydrocarbon bonded stationary phase. The mechanism of retention is subject to debate [7-12], two mechanisms having been proposed. In the partition mode in the ionic or ionizable sample molecule, which has insufficient lipophilic character to be retained, forms an ion-pair with a suitable counter-ion (or ion-pair agent) added to the mobile phase. The formation of the ion-pair increases the lipophilic character of the sample and increases its affinity for the stationary phase.

In the ion-exchange mode the polar counter-ion is assumed to be absorbed by the hydrocarbon stationary phase thus creating an ion-exchange site on to which the polar sample molecule is adsorbed much as in ion-exchange chromatography.

However, these two views are certainly an oversimplification, but many of the factors controlling retention may be understood using these hypotheses.

IPC can be applied to the separation of samples containing both ionic and non-ionic compounds. Strong acids (sulphonated dyes) and bases (quaternary amines) are completely ionized in the pH range 2–8 and in the absence of an ion-pairing reagent will be poorly retained on a hydrocarbon stationary phase. However, the ionization of weak acids (amino acids, carboxylic acids) and weak bases (catecholamines) can be controlled by choice of pH in this range so that the equilibrium either lies to the left of the relationship:

$$\text{non-ionic} \rightleftharpoons \text{ionic}$$

in which case they elute as non-ionic components on a non-polar stationary phase, or to the right, in which case they can interact with the ion-pair agent as for strong acids or bases. As well as pH control, the nature and size of the counter-ion and, to a lesser extent, the nature of the solvent and stationary phase all have an effect on the degree of selectivity.

IPC is an alternative to ion-exchange chromatography but offers the advantages of longer column life and greater reproducibility.

1.3.2 Liquid–Solid (Adsorption) Chromatography (LSC)

Liquid–solid chromatography as developed by Tswett in 1906 using 'classical' column chromatography is the original form of liquid chromatography.

The separation is carried out with a liquid mobile phase and a solid stationary phase which reversibly adsorbs the solute molecules. The stationary phase may be either polar (e.g. silica gel, porous glass beads or alumina) when the mobile phase would be relatively non-polar (e.g. hexane or chloroform), or non-polar (e.g. polymer beads) when a polar mobile phase (e.g. water or ethanol) would be used. This latter mode is known as *reverse-phase adsorption*.

The retention of samples in LSC is more predictable than in LLC. since the elution order follows the polarity of the solutes. Furthermore, there is a vast literature on thin-layer chromatography (TLC), the results of which can usually be transferred more or less directly to HPLC. Because of the simplicity and flexibility of TLC it is often more convenient to establish the best combination of mobile and stationary phase using this technique before transferring the separation to HPLC to utilize its greater speed and efficiency. An example of this is illustrated in Fig. 1.1.

A feature of LSC is the degree of selectivity that can be introduced into the technique. The method is less sensitive to molecular weight differences between the solute species than is LLC but it is highly sensitive to compound type. Complex samples can therefore be separated into classes of compounds having the same functional groups, whereas molecules that differ only in alkyl chain length would be poorly separated.

The explanation of this selectivity lies in the nature of the adsorption process. Solvent molecules in the mobile phase compete with the solute molecules for sites on the adsorbent. A set of equilibria is involved:

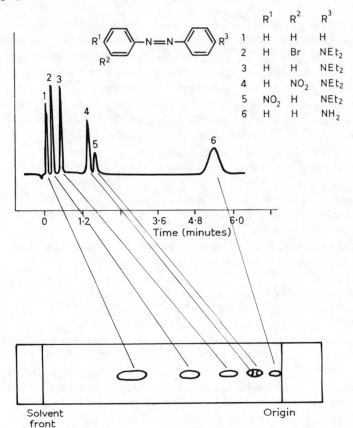

	R^1	R^2	R^3
1	H	H	H
2	H	Br	NEt_2
3	H	H	NEt_2
4	H	NO_2	NEt_2
5	NO_2	H	NEt_2
6	H	H	NH_2

Fig. 1.1 TLC and HPLC of a dyestuff mixture. TLC conditions: silica gel F 254; 10% CH_2Cl_2 in hexane; development time 50 min. HPLC conditions: 15 cm × 2 mm i.d. MicroPak Si 10; 10% CH_2Cl_2 in hexane at 132 cm^3 per hour; 350 lbf in^{-2}; u.v. Detector; 0.2 mg cm^{-3} of each dye (1 μl injected).

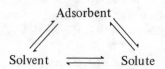

In order that a solute molecule can be adsorbed on to the stationary phase, a solvent molecule must first be displaced from the surface. If it is assumed that the adsorbent possesses a polar surface (e.g. silica or alumina), non-polar groups (e.g. hydrocarbons) will have little affinity for the surface and will not displace the solvent molecules; they will not therefore be retained. Polar functional groups or groups capable of hydrogen-bonding will have a strong affinity for the surface and will be strongly retained. Polarizable molecules (aromatic molecules and high-molecular-weight compounds) will exhibit dipole-induced dipole interactions with the adsorbent surface and will therefore also be retained, the degree of retention depending on the ease of polarization of the functional group or molecule.

For non-polar adsorbents (e.g. charcoal) the dominant intermolecular forces are the London (dispersion) forces, and polar and polarizable molecules will be less strongly retained.

Localized adsorption, or the existence of discrete adsorption sites that are a 'fit' for the adsorbing molecule, also plays an important part in this selectivity, particularly in the separation of isomers. This high level of selectivity is exemplified by the separation of positional isomers, e.g. *m*- and *p*-dibromobenzene. Because of the differences in molecular geometry the *para* isomer is able to interact with two surface −OH groups, whereas the *meta* isomer can only interact with one and is less strongly retained.

Although the new generation of adsorbents are an improvement on those introduced for use in gas–solid chromatography, batch-to-batch reproducibility of adsorbents can still be a problem. Some workers have attempted to standardize batch-to-batch adsorbents by adjusting the water content to give duplicate k' values for a given solute. However, since a single column may, if properly treated, be used for a hundred or more separations, the initial purchase of a 'stock' of adsorbent can minimize the problem of batch-to-batch reproducibility.

Catalytic reactions induced by the adsorbent are not the problem in liquid chromatography as they sometimes are in gas chromatography, because of the lower temperatures employed.

1.3.3 Ion-Exchange Chromatography

Ion-exchange chromatography, which is a form of adsorption chromatography, has been used as a separative technique for over 30 years.

Ion exchange involves the substitution of one ionic species for another. The stationary phase consists of a rigid matrix, the surface of which carries a net positive charge to give an ion-exchange site (R^+). If a mobile phase containing anions is used, the exchange site will attract and hold a negative counter-ion (Y^-). Sample anions (X^-) may then exchange with the counter-ions (Y^-).

The process can be represented in terms of the equilibrium:

$$R^+Y^- + X^- \rightleftharpoons R^+X^- + Y^-$$

Since the process involves the exchange of anions it is known as *anion exchange*. The complementary process of *cation exchange* occurs when the surface carries a net negative charge to give an exchange site (R^-). The counter-ions (Y^+) and the sample ions (X^+) are then both cations and their exchange may be represented by:

$$R^-Y^+ + X^+ \rightleftharpoons R^-X^+ + Y^+$$

The separation is thus based on the strength of the interactions between the sample ions and the exchange site. Ions that interact only weakly with the exchange site will be poorly retained and will have small k' values, whilst ions that have strong interactions will be strongly retained and will have high k' values.

This simple picture of ion exchange can be supplemented by additional processes which may occur under given conditions. The presence of complexing ions may have a marked effect on the process; heavy-metal cations are often complexed with Cl^- ions to give a complex anion and then separated by anion exchange, and the lanthanons are separated on cation exchangers having first complexed the lanthanon cation with citrate ions. Non-ionic species, e.g. sugars, can be separated by ion exchange following the formation of an ionic complex with borate anions [13].

Ligand exchange involves the separation of ligands by virtue of their complexing strength for a metal ion sorbed on the exchange resin, the counter-ion again being displaced. The separation of amino acids on a zinc or cadmium modified resin with ammonium counter-ions is an example of this technique [14].

The existence of one or more of these additional processes makes it difficult to predict the selectvitiy of separations using ion-exchange systems.

Ion exchangers can be further divided into *weak* or *strong*, *anion* or *cation exchangers* according to the nature of the functional groups in the resin. Cation exchangers contain acidic groups (e.g. $-SO_3^-$, strong; or $-COO^-$, weak), whilst anion exchangers contain basic groups (e.g. $-NR_3^+$, strong or $-NH_2$, weak).

The classical ion-exchange resin, e.g. Dowex and Amberlite, is usually a styrene—divinylbenzene copolymer which has been cross-linked to provide mechanical rigidity, with the functional groups built into a matrix. These resins swell when placed in water and, to a lesser extent, in organic solvents, the degree of swelling depending on the degree of cross-linking and the nature of the exchange groups. These materials are also porous, so separation by size can also occur. Furthermore, slow mass transfer effects reduce the efficiency of such columns (300 plates/metre except for microparticle resins).

Chemically bonded exchangers have not the above-mentioned disadvantages of conventional exchange resins but they possess other disadvantages. These exchangers consist of a non-porous silica matrix to which the functional groups are bonded by a covalently bonded cross-linked silicone network:

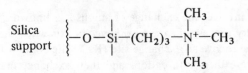

Efficiences up to 25 000 plates per metre are obtained on microparticulate packings and the exchangers do not swell in solvents. However, these exchangers have limited ion-exchange capacities and some can be used only over a limited pH range. The relative merits of the two types of ion exchangers will be discussed in greater detail in Chapter 4.

Ion exchange rarely causes the degradation of a sample. If, however, it should occur, a change to a lower temperature or a change in pH or ionic strength of the carrier is usually sufficient to prevent degradation. Failing this, a change from a strong to a weak ion exchanger may be necessary.

Deactivation of the exchange resin may occur, either by the preferential adsorption of retained sample components, or by the physical blocking of the exchange sites by particulate matter.

The effect of temperature on the selectivity of the resin is difficult to predict, and this factor is usually disregarded in selecting the operating temperature; however, the stability of the exchange resin may be affected by temperature, and the manufacturers' recommendations should be followed. Higher temperatures are an advantage because the decrease in viscosity of the mobile phase with increase in temperature leads to improved mass transfer of the sample components with the consequent increase in column efficiency. Higher temperatures also usually lead to shorter analysis times.

Batch-to-batch reproducibility of ion-exchange resins is often poor and some form of standardization is desirable.

1.3.4   Exclusion Chromatography

Exclusion chromatography, also referred to as *gel filtration, gel permeation chromatography*, or *gel chromatography*, dates from 1959 with the introduction of a dextran gel in bead form marketed under the name of Sephadex. In modern exclusion chromatography a wide range of stationary phases is available and these can be divided into three classes: (i) the aerogels (porous glass or silica); (ii) the xerogels, e.g. cross-linked dextran and polyacrylamide; (iii) xerogel-aerogels, e.g. cross-linked agarose, polystyrene, and polyvinylacetate.

Exclusion chromatography separates substances according to their molecular size and shape. Small molecules that can enter freely into the pores of the stationary phase are said to have a distribution coefficient $K = 1$, and large molecules which are completely excluded from all pores have a distribution coefficient $K = 0$, whilst molecules of an intermediate size will have distribution coefficients between 0 and 1. Thus large molecules will move more rapidly through the column than will the smaller molecules and they are eluted first. Molecules are therefore eluted in order of decreasing molecular size. The process is shown diagrammatically in Fig. 1.2.

Since the solvent molecules are usually very much smaller than the molecules being separated by this method, they are eluted last (at t_m). Hence, contrary to other forms of chromatography, the sample is eluted before t_m.

The mechanism of separation in exclusion chromatography is complex, but the overriding mechanism appears to be that of steric exclusion [15]. Diffusion [16-18] may also play some part, as may adsorption on to the gel, but in properly designed systems the effect of these should be minimal.

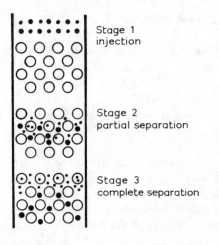

Stage 1
injection

Stage 2
partial separation

Stage 3
complete separation

Fig. 1.2 Schematic diagram of three stages in a separation by exclusion chromatography. Empty circles represent the gel bed, filled circles the solute molecules.

Exclusion chromatography possesses several advantages in its application. Because all of the sample is eluted in a relatively short elution time (before t_m), gradient elution facilities are not required. The short elution times also lead to narrow solute bands which are easier to detect and do not give rise to the problems of detection limits often encountered in other forms of liquid chromatography where the solute bands may become more diffuse. Hence, the less sensitive refractometer detector is usually used in exclusion chromatography.

The retention time is a function of molecular size and can be predicted for a compound of known molecular structure. The beginning of the chromatogram is therefore determined by the retention time predicted for the compound of largest molecular size, and the end by the solvent retention time (t_m). A series of different samples can therefore be injected at predetermined intervals without fear of the chromatograms overlapping each other, and automatic injection systems are widely used in exclusion chromatography. The correlation of retention times with molecular structure also aids the identification of unknown substances.

Because intermolecular forces are absent in the separation process, sample loss or chemical interaction is kept to a minimum and the column does not accumulate strongly retained molecules. This results in a longer than usual column life, and pre-columns are not normally required for clean-up.

The main disadvantages in exclusion chromatography are its inability to resolve compounds with similar molecular size distributions and its limited peak capacity. Molecular weight differences of 10% or more are usually required to achieve separation, so the technique does not lend itself to complete separations of complex mixtures but is used mainly as a preliminary separation method or as a method of obtaining the molecular weight distribution of a polydisperse polymer.

Limited peak capacity arises from the relatively short time required to run the chromatogram (maximum time = t_m), and chromatograms with more than ten peaks are rarely seen.

1.4 Scope of the Techniques

Many problems in chromatography can be solved by more than one method. Others may only be amenable to one particular method. The following is therefore intended only as a general guide; for specific problems the liquid chromatography literature should be surveyed.

1.4.1 Liquid–Liquid Chromatography

Because of the variety of stationary phases available, LLC is the most versatile mode of liquid chromatography, and a wide variety of sample types, both polar and non-polar, can be separated. Separation occurs on the basis of the nature and number of the substituent groups, and on differences in molecular weight.

Some classes of compounds are best separated by normal phase LLC and some by reverse phase LLC. Examples include:

Normal phase LLC — plasticizers, dyes, pesticides, steroids, anilines, alkaloids, glycols, alcohols, phenols, aromatics, and metal complexes.

Reverse phase LLC — alcohols, aromatics, anthraquinones, alkaloids, oligomers, antibiotics, barbiturates, steroids, chlorinated pesticides, and vitamins.

Ion-pair liquid–liquid chromatography has been applied to the separation of biogenic amines, sulphonamides, carboxylates, and sulphonates.

1.4.2 Liquid–Solid Chromatography

LSC is most successfully applied to non-ionic samples of intermediate molecular weight (200 to 2000) which are soluble in organic solvents. Ionic compounds usually give rise to 'tailing' peaks, because of the high surface energies of the adsorbents, unless the surface is deactivated in some way. One way of deactivating the surface is to use water, and it is for this reason that the separation of water-soluble molecules is not usually very satisfactory. Water-soluble samples are better separated using reverse-phase partition chromatography or bonded stationary phases.

Compounds differing in chemical type or in the number of functional groups are well separated by LSC; examples include antioxidants, dyes, vitamins, steroids, barbiturates, amines, hydrocarbons, phenols, alkaloids, amides, lipids, fatty acids and alcohols.

1.4.3 Ion-Exchange Chromatography

Ion-exchange chromatography is mainly used in the analysis of ionized or ionizable compounds, especially in biochemical applications. Many biochemical components (amino acids, nucleic acids, and proteins), which are either too involatile or too labile for separation by GC, may be ionized by suitable choice of the pH of the solution and separated by ion-exchange chromatography.

However, the technique may also be applied to a wide range of separations including carboxylic acids, aromatic sulphonates, sugars, analgesics, vitamins, purines, glycosides, and inorganic anions and metal cations.

1.4.4 Exclusion Chromatography

Although the literature contains examples of the application of exclusion chromatography to the separation of both organic and inorganic molecules in aqueous and non-aqueous systems, the technique is predominantly used for the analysis of high-molecular-weight compounds (>2000) including organic polymers (e.g. polyolefins, polystyrene, polyvinyls and polyamides), silicones, and biopolymers (e.g. proteins, nucleic acids, oligosaccharides, peptides, sugars and glycols).

Exclusion chromatography is also finding increasing application in the characterization of high polymers, by the determination of their molecular weight distribution [19].

References

1. James, A. T. and Martin, A. J. P. (1952) *Biochem. J.*, **50**, 679.
2. Huber, J. F. K. and Hulsman, Y. A. R. J. (1967) *Anal. Chim. Acta*, 38, 405.
3. Giddings, J. C. (1965) *Dynamics of Chromatography*, Marcel Dekker, New York.
4. Martin, A. J. P. and Synge, R. L. M. (1941) *Biochem. J.*, **35**, 1358.
5. Halasz, I. and Sebastian, I. (1969) *Angew. Chemie*, 81, 464.
6. Eksborg, S., Lagerström, P., Modin, R. and Schill, G. (1973) *J. Chromatog.*, 83, 99.
7. Kissinger, P. T. (1977) *Anal. Chem.*, **49**, 883.
8. Horvath, C., Melander, W., Molnar, I. and Molnar, P. (1977) *Anal. Chem.*, **49**, 2295.
9. Bidlingmeyer, B. A., Deming, S. N., Price, Jr. W.P., Sachok, B. and Petrusek, M. (1979) *Advances in Chromatography* (ed. A. Zlatkis), Houston, U.S.A., p. 435.
10. Konijnendijk, A. P. and van de Venne, J. L. M. (1979) *Advances in Chromatography*, (ed. A. Zlatkis), Houston, U.S.A., p. 451.
11. Jansson, S. O., Andersson, I. and Persson, B. A. (1981) *J. Chromatog.*, **203**, 93.
12. Knox, J. H. and Hartwick, R. A. (1981) *J. Chromatog.*, **204**, 3.
13. Khym, J. X., Zill, L. P. and Cohn, W. E. (1957) *Ion Exchangers in Organic Biochemistry* (ed. C. Colman and T. R. E. Kressman), Interscience, New York.
14. Arikawa, Y. and Makimo, I. (1966) *Fed. Proc.*, **25**, 786.
15. Flodin, P. (1962) in *Dextran Gels and Their Applications in Gel Filtration*, Pharmacia, Uppsala, Sweden.
16. Ackers, G. K. (1964) *Biochemistry*, **3**, 723.
17. Yau, W. W. and Malone, D. P. (1967) *J. Polymer Sci.*, B, **5**, 663.
18. Di Marzio, E. Z. and Guttman, C. M. (1969) *J. Polymer Sci.*, **7**, 267.
19. Billingham, N. C. (1976) in *Practical High Performance Liquid Chromatography* (ed. C. F. Simpson), p. 167, Heyden, London.

2

Chromatographic theory

2.1 The Process of Separation

Chromatography involves the separation of the components of a mixture by virtue of differences in the equilibrium distribution (K) of the components between two phases: the mobile phase and the stationary phase. If C_s and C_m are the concentrations of a component in the stationary and mobile phases respectively, then:

$$K = C_s/C_m \qquad (2.1)$$

Migration of component molecules may be assumed to occur only when the molecules are in the mobile phase. The rate of migration of a component is then inversely proportional to its distribution coefficient, so components with a high distribution in the stationary phase will move more slowly through the column and hence be separated from the components with a lower distribution in the stationary phase. Without this difference in distribution and, by inference, a differential rate of migration, no separation can be achieved.

Differential migration therefore depends upon the experimental variables that affect the distribution, i.e. the composition of the mobile and stationary phases and the temperature. The effect of the column pressure on the distribution coefficients is negligible at the pressures usually used in liquid chromatography.

2.2 Retention in Liquid Chromatography

Owing to retention, chromatographic zones move through the column at a rate less than the mobile phase velocity. The ratio of the two velocities is known as the retardation factor (R):

$$R = \frac{\text{rate of movement of the sample band}}{\text{rate of movement of the mobile phase}} \qquad (2.2)$$

The time of elution of the peak maximum, called the *retention time* (t_R), can be related to the equilibrium distribution coefficient. Fig. 2.1 shows a simple separation of a two-component mixture eluting with retention times t_{R1} and t_{R2}.

The elution time is a function of the mobile phase velocity, and the volume of mobile phase required to elute a component from the column, the *retention volume* (V_R), is given by:

$$V_R = F \times t_R \tag{2.3}$$

where F is the volume flow rate of the mobile phase.

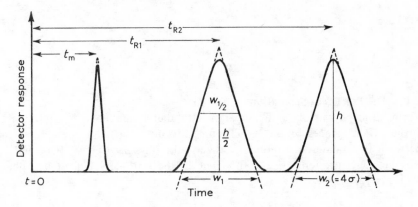

Fig. 2.1 Separation of a two component mixture.

Fig. 2.1 also illustrates the elution of an 'unretained component', i.e. one that has a distribution coefficient in a particular stationary phase of zero, so that it is either non-adsorbed or is insoluble. In the absence of exclusion effects it will therefore pass through the column at the same rate as the mobile phase with a retention time t_m*. Large molecules could be excluded from all or part of the pore structure of the stationary phase and they would then elute in a volume less than the mobile phase volume as in exclusion chromatography.

The volume of the column available to the mobile phase is:

$$V_m = F \times t_m \tag{2.4}$$

This mobile phase volume (or dead volume) contributes nothing to the separation and is dependent on the geometry of the column and the packing. A contribution to the mobile phase volume will also be made by the injection system and by the detector and its pipework. Thus the contributions may be separated into 'in-column' and 'extra-column' effects. In gas chromatography extra-column effects are negligible but in liquid chromatography they may be considerable. For this reason special attention must be paid to the design of injection systems and detectors to minimize dead volumes.

* It has been the practice to use t_0 for the retention time of the unretained peak. We prefer to use t_0 to indicate the start of a chromatogram (i.e. the injection) at time $t = 0$, and to use t_m, since it is related to the mobile phase velocity, for the retention of an unretained peak.

In liquid chromatography the mobile phase may be considered to be incompressible, and the pressure and flow rate are constant throughout the column. In contrast to gas chromatography, therefore, no compressibility correction has to be made to flow measurements taken at the column outlet to obtain the mean flow through the column.

In simple chromatographic theory the occurrence of an elution peak with a Gaussian distribution in the chromatogram is taken as showing a linear relationship between the concentration of the sample molecules in the stationary and mobile phases; i.e. the distribution coefficient (K) is a constant and the isotherm is linear. Under these conditions the retention time is independent of the sample size. Skewed peaks are the result of non-linear distribution isotherms, as shown in Fig. 2.2, and the retention time will vary with sample concentration.

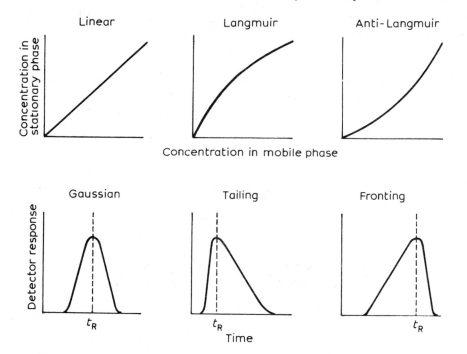

Fig. 2.2 Three basic isotherm shapes and their effect on peak shape and retention time.

If the sample is sufficiently small it should be noted that symmetrical Gaussian peaks can be obtained even for components with a non-linear isotherm, and retention times will again be independent of the sample size.

The fundamental retention equation for partition chromatography, neglecting non-linearity and band broadening, is:

$$V_R = V_m + KV_s \qquad (2.5)$$

where V_R is the retention volume, V_m is the volume of the mobile phase, V_s is the volume of the stationary phase, and K is the equilibrium distribution coefficient.

In adsorption chromatography the volume of the stationary phase is replaced by the surface area of the adsorbent (A_s) so that:

$$V_R = V_m + KA_s \tag{2.6}$$

Since V_m plays no part in the separation process and is also constant for all components of a mixture, the net retention volume, V_N, is often used:

$$V_N = V_R - V_m \tag{2.7}$$

or

$$V_N = KV_s \tag{2.8}$$

In adsorption chromatography the net retention volume is sometimes normalized for the mass of the adsorbent on the column (W_a) so that:

$$V_N{}^\circ = V_N/W_a \tag{2.9}$$

If several independent retention mechanisms are contributing to the separation, as in ion-exchange chromatography for example, the overall net retention volume is the sum of the contributions of the retention volumes from the various mechanisms.

A fundamental chromatographic parameter is the capacity factor k' where:

$$k' = n_s/n_m \tag{2.10}$$

and n_s and n_m are the number of moles of solute in the stationary and mobile phases respectively.

Since a solute molecule migrates down the column only when it is in the mobile phase, the retardation factor R may be written [1]:

$$R = \frac{\text{amount of solute in mobile phase}}{\text{amount of solute in mobile + stationary phases}} \tag{2.11}$$

$$= \frac{n_m}{n_m + n_s} \tag{2.12}$$

$$= \frac{1}{1 + k'} \tag{2.13}$$

If K is the equilibrium distribution coefficient, then:

$$k' = K(V_s/V_m) \tag{2.14}$$

Combination of Equations (2.5) and (2.14) gives:

$$V_R = V_m(1 + k') \tag{2.15}$$

When $k' = 0$ it can be seen that:

$$V_R = V_m$$

and no separation will take place.

If the average linear velocity of the chromatographic zone is v and the zone traverses the column of length L (cm) in time t_R (s), then:

$$t_R = L/v \qquad (2.16)$$

Similarly, the time t_m for solvent molecules (or an unretained compound) to traverse the column is:

$$t_m = L/v_m \qquad (2.17)$$

where v_m is the mobile phase velocity.

Equation (2.15) can be written in terms of the retention times t_R and t_m:

$$t_R = t_m(1 + k') \qquad (2.18)$$

or on rearrangement:

$$k' = \frac{t_R - t_m}{t_m} \qquad (2.19)$$

Substitution of Equation (2.17) in Equation (2.18) gives:

$$t_R = \frac{L}{v_m}(1 + k') \qquad (2.20)$$

Hence the retention time t_R is directly proportional to the column length and inversely proportional to the linear flow rate of the mobile phase.

Equation (2.19) gives a direct means of determining the k' values for the components of a mixture.

The parameter t_m can be obtained in one of several different ways. If a molecule that has 'weaker' solvent properties is injected as a sample, its t_R value will be equal to t_m, as is the case when a sample of pentane is injected into a hexane mobile phase. A second method depends on the fact that equilibrium constants, and hence retention times, will vary with temperature, If, therefore, there is no change in the t_m value when the temperature is varied by say 20°C, it may be assumed that the molecule is not retained. A third method, involving the use of radioactively labelled solvent molecules and a radiation detector, is rarely used.

2.3 Band Broadening – Origins

As a solute band passes through the column, the width of the band increases and the solute is diluted by the mobile phase. There are three main contributions to band broadening: eddy diffusion, molecular diffusion, and mass transfer.

The relative importance of band broadening in liquid chromatography compared with that in gas chromatography is explicable in terms of the relative order of magnitude values of the physical constants for liquids and gases given in Table 2.1 [2].

Table 2.1   *Order of magnitude values of physical properties important in determining the extent of band broadening*

Property	Gas	Liquid
Diffusion coefficient, D_m ($cm^2\ s^{-1}$)	10^{-1}	10^{-5}
Density, ρ ($g\ cm^{-3}$)	10^{-3}	1
Viscosity, η ($g\ cm^{-1}\ s^{-1}$)	10^{-4}	10^{-2}
Reynolds number	10	100

Hence, in gas chromatography, eddy diffusion, molecular diffusion, and mass transfer effects on the stationary phase are important, whilst in liquid chromatography, molecular diffusion can usually be ignored but mass transfer effects in the mobile phase as well as in the stationary phase become important.

2.3.1   Eddy Diffusion
The flow pattern through a bed of granular particles is very tortuous as the molecules take the line of least resistance to fluid flow. Fluids move more slowly in narrow flow paths and more rapidly in wider ones. A single molecule therefore may have its velocity through the particle bed varied between wide limits. Because of the random nature of the packing, these velocity changes will also be random. The simple theory of eddy diffusion assumes that the solute molecules are fixed in given flow paths. In practice, however, this is not true. The solute molecules may diffuse laterally from one flow path into another where the flow velocity may be quite different. This 'coupling' of lateral diffusion with the normal process of eddy diffusion results in a decrease in the amount of band broadening and hence an increase in efficiency of the chromatographic process. Giddings [1] has given a full account of the dynamics of flow spreading.

2.3.2   Molecular Diffusion
Diffusion in the direction of flow (longitudinal molecular diffusion) in the fluid phase, and to a much lesser extent at interfaces between phases, e.g. on solid surfaces, is an important band-broadening process in gas chromatography, but because of the lower rates of diffusion in liquids as compared with gases this process may usually be ignored in liquid chromatography.

2.3.3   Mass Transfer
Mass transfer effects may be conveniently divided into stationary phase and mobile phase mass transfer terms.

Stationary phase mass transfer (Fig. 2.3). The rate at which solute molecules transfer into and out of the stationary phase makes a significant contribution to band broadening and hence to efficiency. This rate depends mainly on diffusion for liquid stationary phases and on adsorption—desorption kinetics for solid stationary phases. Solute molecules will reside in or on the

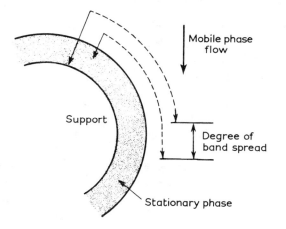

Fig. 2.3 Stationary phase mass transfer.

stationary phase for varying lengths of time; e.g. with a liquid stationary phase some molecules will diffuse more deeply into the liquid layer. When these molecules are desorbed they will have been left behind, the bulk of the molecules and the solute band will have been broadened. Likewise, in adsorption chromatography those solute molecules adsorbed on active sites will have a larger residence time on the surface and will again be left behind the main stream of molecules when they are finally desorbed. Hence, to reduce band broadening from stationary phase mass transfer effects, liquid layers should be as thin as is possible without introducing adsorption effects on the support material, and solid surfaces should be of a homogeneous nature.

Mobile Phase Mass Transfer. Mobile phase mass transfer effects in LC must be divided into contributions from (a) the 'moving' mobile phase and (b) the 'stagnant' mobile phase.

(a)   Moving mobile phase mass transfer (Fig. 2.4). Molecules in the *same* flow path will not all move with the same speed. Those molecules close to the particle walls will move more slowly than those in mid-stream, and a flow profile will develop across the channel and band broadening will increase.

(b)   Stagnant mobile phase mass transfer (Fig. 2.5). When porous stationary phases are used, the intraparticle void volume is filled with mobile phase at rest. Solute molecules must diffuse through this stagnant mobile phase in order to reach the stationary phase. Molecules that diffuse only a short distance into the pore will rapidly regain the mainstream, whereas molecules that diffuse further

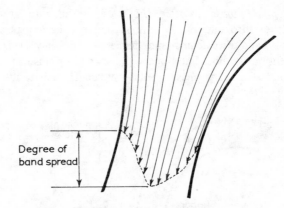

Fig. 2.4 'Moving' mobile phase mass transfer.

and spend more time in the pore will be left behind the mainstream, again resulting in broadening of the chromatographic band.

The adverse effect of stagnant mobile phase mass transfer has been instrumental in the development of specialized support materials for use in liquid chromatography.

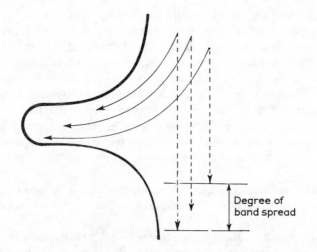

Fig. 2.5 'Stagnant' mobile phase mass transfer.

2.4 Band Broadening and the Plate Height Equation

The *efficiency* or *number of theoretical plates* N of a chromatographic system may be defined from a single chromatographic band as in Fig. 2.1.

$$N = 16(t_R/w)^2 \qquad (2.21)$$

where t_R is the uncorrected retention time and w is the peak width at the base line measured in units of time.

An equivalent expression which avoids the problem in measuring the basal peak width is:

$$N = 5.54(t_R/w_{\frac{1}{2}})^2 \qquad (2.22)$$

where $w_{\frac{1}{2}}$ is the peak width at half peak height.

Since the band width increases as the band proceeds down the column (i.e. as t_R increases), N is a measure of the relative band broadening with distance down the column.

For a given set of operating conditions the quantity N is approximately constant for different bands in the chromatogram and is therefore a measure of the *column efficiency*. In comparing column efficiencies a more useful parameter is the *height equivalent to a theoretical plate (HETP)* or *plate value* H, where:

$$H = L/N \qquad (2.23)$$

L is the length of the column and H measures the efficiency of the column per unit length. Small H values (large N values) therefore mean more efficient columns and this is one of the main goals in chromatography.

Each of the band broadening processes discussed in the preceding section contributes to the overall plate value for the column. Their magnitude and relative importance, however, depend upon the particular chromatographic technique under consideration, and even within a technique their importance may change with changing operating conditions. For example, as already mentioned, longitudinal molecular diffusion may play little part in liquid chromatography, and in gel chromatography, where there is in principle no sorption, band broadening arises only from mobile phase effects.

The overall plate value can be represented as the sum of the contributions from longitudinal diffusion (H_L), stationary phase mass transfer (H_S), 'moving' mobile phase effects (H_M), and 'stagnant' mobile phase effects (H_{SM}).

2.4.1 Longitudinal Diffusion (H_L)

The plate height contribution of longitudinal diffusion H_L is given by:

$$H_L = 2\gamma D_M/v \qquad (2.24)$$

where γ is an obstruction factor which accounts for the fact that longitudinal diffusion is hindered by the column packing. In packed columns its value is about 0.6. In liquid chromatography, because the solute diffusion coefficient in the mobile phase D_M is small, the contribution of H_L to the overall plate height is also small and in most cases it may be assumed to be zero. In this case the H versus v relationship will not show a minimum.

2.4.2 Stationary Phase Mass Transfer (H_S)

The one-dimensional random walk model of Giddings [1] gives the plate height contribution for a first order adsorption–desorption process as:

$$H_S = 2\left(\frac{k'}{1 + k'}\right)^2 vt_a \qquad (2.25)$$

where t_a is the time spent by a molecule in the mobile phase before adsorption occurs.

It is often more convenient to express H_S in terms of the mean desorption time t_d (the mean time that a molecule remains attached to the surface).

The ratio t_a/t_d is therefore the ratio of times spent in the mobile and stationary phases, and therefore:

$$t_a/t_d = 1/k' \qquad (2.26)$$

and

$$H_S = 2vt_d \frac{k'}{(1 + k')^2} \qquad (2.27)$$

Thus the smallest contribution to the plate height occurs at low mobile phase velocities and with rapid mass transfer. The capacity factor function will show a maximum at $k' = 1$ and will then decrease with increasing k' value.

Mass transfer in a bulk stationary phase, as in partition systems, is analogous to retention on an adsorptive surface in that a certain average time is required to absorb and desorb the molecule. However, whereas the mean desorption time is determined by the rate constant (k_d) for the desorption process, in a partition system the controlling factor is the solute diffusion coefficient in the stationary phase D_S. The mean desorption time is replaced by the average diffusion time t_D, the time taken for a molecule to diffuse a distance d in the liquid. Then:

$$t_D = d^2/2D_S \qquad (2.28)$$

and the plate height contribution becomes:

$$H_S = \frac{k'}{(1 + k')^2} \cdot \frac{d^2 v}{D_S} \qquad (2.29)$$

The distance d may be equated approximately with the thickness of the stationary liquid film d_f so that slow mass transfer is equated with high liquid loadings. It is also desirable to choose liquids within which the solute molecule has a large diffusion coefficient D_S.

In a rigorous treatment of diffusion controlled mass transfer the precise shape of the partitioning liquid should be considered. Equation (2.29) should thus be multiplied by a *configuration* factor q. For a uniform liquid film $q = 2/3$. For diffusion in a rod shaped pore (as in paper chromatography) or in a spherically shaped body (an ion-exchange bead) the values of q are $1/2$ and $2/15$ respectively.

2.4.3 'Moving' Mobile Phase Effects (H_M)

The flow of a fluid through a packed bed is too complicated for a rigorous treatment, involving as it does flow inequalities due to eddy diffusion and lateral mass transport by diffusion and by convection. Giddings [1] has shown that the plate height contribution from mobile phase effects H_M can be related to contributions from eddy diffusion (H_F) and diffusion (H_D). The contribution from eddy diffusion is:

$$H_F = 2\lambda d_p \tag{2.30}$$

where d_p is the particle diameter and λ is a packing constant.

The diffusion term H_D may be written:

$$H_D = \Omega v d_p^2 / D_M \tag{2.31}$$

where, again, Ω is a function of the packing structure.

We have already noticed that the eddy diffusion may be coupled with lateral diffusion. It was once assumed that H_M could be taken to be the sum of the two effects:

$$H_M = H_F + H_D \tag{2.32}$$

However, a study of the quantitative aspects of the combined flow-diffusive exchange leads to the relationship:

$$H_m = \frac{1}{1/H_F + 1/H_D} = \frac{1}{1/2\lambda d_p + D_M/\Omega v d_p^2} \tag{2.33}$$

The differences between the classical form $H = H_F + H_D$ and the above equation are exemplified in Fig. 2.6.

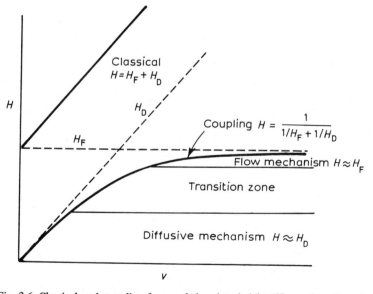

Fig. 2.6 Classical and coupling forms of the plate height (H) as a function of velocity (v) (after J. C. Giddings).

On the coupling theory the value of H is always less than H_D or H_F, and very much less than the sum $H_F + H_D$. At high mobile phase velocities, H_M approaches the constant value H_F, and at low velocities it approaches H_D. In other words, at low mobile phase velocities lateral diffusion is the controlling factor whilst at high velocities eddy diffusion is most important.

In general H_M increases with particle diameter and flow velocity, and decreases with solute diffusivity.

2.4.4 'Stagnant' Mobile Phase (H_{SM})

The plate height contribution from mobile phase trapped as 'stagnant' mobile phase in porous spherical particles is:

$$H_{SM} = \frac{(1 - \phi + k')^2 d_p{}^2 v}{30(1 - \phi)(1 + k')^2 \gamma D_M} \qquad (2.34)$$

where ϕ is the fraction of total mobile phase in the intraparticle space and λ is a tortuosity factor inside the particle.

In exclusion chromatography mobile phase effects are the only contribution to band broadening since in theory no sorption occurs. The plate height contribution for the stagnant mobile phase is then found to be:

$$H_{SM(G)} = \frac{k'}{30(1 + k')^2} \cdot \frac{d_p{}^2 v}{D_M} \qquad (2.35)$$

2.5 Overall Plate Height Equation

The overall plate height equation may be obtained by summing the contributions from the various band-broadening processes. Thus:

$$H = H_L + H_S + H_M + H_{SM} \qquad (2.36)$$

In order to optimize for a minimum plate height it is convenient to write the individual contributions in terms of those physical quantities which are most easily controlled, e.g. d_f, d_p, D_M and D_S, thus:

$$H_L = C_D{}'(D_M/v) \qquad (2.37)$$

$$H_S = C_S{}'(d_f{}^2 v/D_S) \qquad (2.38)$$

$$H_F = C_F{}'d_p \qquad (2.39)$$

$$H_D = C_D{}'(d_p{}^2/D_M) \qquad (2.40)$$

$$H_{SM} = C_{SM}{}'(d_p{}^2 v/D_M) \qquad (2.41)$$

from which:

$$H = C_D{}'\frac{D_M}{v} + C_S{}'\frac{d_f{}^2 v}{D_S} + C_{SM}{}'\frac{d_p{}^2 v}{D_M} + \frac{1}{1/C_F{}'d_p + D_M/C_M{}'vd_p{}^2} \qquad (2.42)$$

As already noted, H_L may often be ignored in liquid chromatography.

In order to minimize H, therefore, it is necessary to use columns packed with small diameter particles, a slow-moving solvent of low viscosity, and high separation temperatures, to increase diffusion rates.

2.6 Comparison with Gas Chromatography

The variation of H with v in gas chromatography is represented by the simplified form of the van Deemter equation:

$$H = A + B/v + Cv \qquad (2.43)$$

where the terms A, B, and C represent plate height contributions from eddy diffusion, longitudinal molecular diffusion, and mass transfer effects respectively.

Equation (2.42) may be written in the same form:

$$H = B/v + C_S v + C_{SM} v + (1/A + 1/C_M v)^{-1} \qquad (2.44)$$

or, neglecting longitudinal molecular diffusion:

$$H = C_S v + C_{SM} v + (1/A + 1/C_M v)^{-1} \qquad (2.45)$$

Fig. 2.7 shows a typical H versus v plot in GC and in LC.

The H versus v curve for GC shows a minumum value of the plate height H_{min} at mobile phase velocity v_{min}. Below this optimum velocity, H is dependent on the B term of the van Deemter equation, i.e. longitudinal

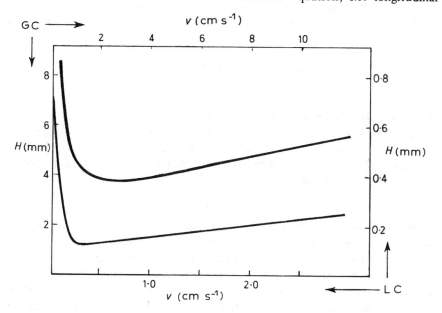

Fig. 2.7 Typical H versus v curves in gas chromatography and liquid chromatography.

molecular diffusion is the controlling process. At higher flow rates the mass transfer term C is the controlling factor. The H versus v curve for LC rarely shows a minimum because the B term normally plays no practical role in LC and the minimum when observed occurs at very low flow velocities [3] ($\sim$0.1 mm s^{-1}). However, as shown by Halasz *et al.* [4], if the particle size is less than 5 μm, a minimum in the H versus v curve is observed and at velocities that are significant in routine analytical analysis.

The LC curve also shows a flatter rise of H with velocity than in GC. This is due to the complex moving mobile phase flow process which results in the coupling of H_F and H_D. At high flow velocities the C_M term approaches the simple C_F term. The flat slope of the H versus v curve in LC means that high mobile phase velocities can be used without a serious loss in column efficiency.

The dependence of H on v can also be represented by the empirical equation [5]:

$$H = Dv^n \tag{2.46}$$

where D and n are characteristic constants for a given column which can vary slightly with other experimental conditions. The constant D is equal to H for $v = 1$ cm s^{-1}, and D is therefore an approximate measure of the column efficiency. The constant n usually falls within the limits $0.3 \leqslant n \leqslant 0.7$ although it does occasionally exceed these limits. Porous packings generally give values of $n = 0.6$ and pellicular packings $n = 0.4$. Low values of n are indicative of poorly packed columns, and high values of slow stationary phase mass transfer. For example, porous ion exchangers show values of n close to unity, and also have much larger H values than porous packings of other types.

The validity of Equation (2.46) has been confirmed for a wide variety of LC systems involving 10–100 μm particles and values of v between 0.5 and 10 cm s^{-1}.

An extension of Equation (2.46) enables the chromatographer to optimize the resolution in LC [6].

2.7 Column Efficiency and Particle Diameter

Equation (2.42) shows that efficiency tends to increase as smaller diameter particles are used, whereas Equation (2.66) shows that the permeability decreases ($K° \propto d_p^2$). The dependence of plate height on particle diameter has been found to follow the relationship:

$$H \approx d_p^\beta \tag{2.47}$$

where the value of β becomes increasingly smaller as the particle size decreases ($\beta = 1.8$–1.3). Halasz *et al.* [4] have shown that, when the particle diameter is less than 5 μm, $H \approx d_p$, and that the minimum particle size in HPLC is between 1 and 3 μm. However, with particles less than 10 μm, efficient packing becomes increasingly difficult as the particle diameter decreases, but as long as the column is 'well packed' smaller particles give more efficient columns regardless of column pressure or separation time.

Equation (2.47) fails to predict the increase in plate height at very low flow rates so that its application is limited.

2.8 Reduced Plate Height and Reduced Velocity

The use of 'reduced parameters' is to be preferred for some uses [1,4]. The 'reduced plate height' h and the 'reduced velocity' v are defined by:

$$h = H/d_p \tag{2.48}$$

and

$$v = vd_p/D_m \tag{2.49}$$

where d_p is the particle diameter and D_m is the solute diffusion coefficient in the mobile phase.

Since both reduced parameters are normalized for the particle diameter, it is to be expected that reduced height-reduced velocity curves for columns packed with different particle diameters should be similar. This has been confirmed in practice and is justification for the use of reduced parameters.

For an efficient column, h is around 2–3, and a value of h greater than 10 may be taken as an indication of a poorly packed column or of poor column materials.

Although in gas chromatography the reduced velocity is not much higher than unity, in liquid chromatography the small D_m values mean that the v values can be up to several thousands.

In terms of the reduced plate height, an analogous equation to Equation (2.43) may be written [7]:

$$h = Av^{0.33} + B/v + Cv \tag{2.50}$$

where $Av^{0.33}$ represents the contribution from eddy diffusion, B/v the contribution from longitudinal molecular diffusion, and Cv the contribution from the mass transfer terms.

The way in which h depends on v is shown in Fig. 2.8.

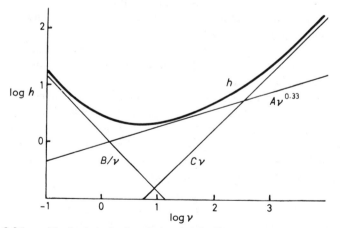

Fig. 2.8 Logarithmic plot of reduced plate height (h) versus reduced velocity (v), showing contributions from the three terms in Equation (2.47) (after J. H. Knox).

From the curve the reduced velocity corresponding to the minimum reduced plate height may be obtained. This corresponds to a value ≈ 3. It is then possible to calculate the particle diameter necessary to give this reduced velocity. This is typically ~ 5 μm and, preparative applications apart, there is no justification for using particles with diameters greater than 10 μm.

2.9 Extra-Column Band Broadening

Band broadening other than in the column is negligible in gas chromatography. However, in LC extra-column band broadening contributes a variance to H which in general leads to a lowering of column efficiency. This extra-column band broadening can occur in the injection system, in connecting pipework, and in the detector, and good instrument design should aim at keeping volumes in these three regions of the liquid chromatograph to a minimum.

As long as band widths are measured in volume units, the band broadening occurring in different regions of the chromatogram is additive. The total band broadening should be less than 30% of the volume of the narrowest band in the chromatogram since then the increase in the band width will be limited to 10% or less.

Connecting tubing between the column and the detector may be the major factor in band broadening [8]. The band broadening σ_v occurring in a tube of length L and radius r is given by [9]:

$$\sigma_v^2 = \pi r^4 \, FL/24 D_M \qquad (2.51)$$

where F is the volume flow rate and D_M is the diffusivity. It is necessary therefore to have short lengths of narrow bore tubing. Tubing with an internal diameter of about 0.01 inch is often used; narrower bore tubing tends to obstruct, and wider bore tubing contributes further to band broadening. The maximum length of tubing used should not exceed 30 cm.

If the detector is not to contribute to band broadening, the detector cell volume should not exceed 10 μl, but where high efficiency columns are being used a still smaller volume is desirable.

For a given column the influence of extra-column effects depends on the k' value of the solute, and these effects are most noticeable at low k' values. A critical test of extra-column effects is to study the H versus v relationship for an unretained peak ($k' = 0$). If high H values are observed, extra-column effects are probably a major contributory factor. Extra-column effects also become more important when short columns are being used.

Sample introduction may also lead to an increase in band broadening. Sample introduction is best achieved by on-column injection through a septum or by stop-flow injection. The use of sample valves may contribute to band broadening because of slow removal of solute molecules from the walls of the valve.

2.10 Resolution

Substances are separated in a chromatographic column when their rates of migration differ. The ability of a particular stationary phase (or solvent) to produce the separation is a function of the thermodynamics of the system. We can quantify this *solvent efficiency* in terms of the *relative retention* (α) for two components:

$$\alpha = \frac{t'_{R_2}}{t'_{R_1}} = \frac{t_{R_2}-t_m}{t_{R_1}-t_m} = \frac{k'_2}{k'_1} = \frac{K_2}{K_1} \tag{2.52}$$

The relationship $\alpha = K_2/K_1$ emphasizes the thermodynamic nature of α.

More fundamentally it can be shown that:

$$\Delta(\Delta G^{\ominus}) = -RT \ln \alpha \tag{2.53}$$

where $\Delta(\Delta G^{\ominus})$ is the difference in free energies of distribution of the two components.

Having achieved a separation it is necessary to prevent remixing of the components, and the ability to achieve this is a function of the column geometry. This *column efficiency* is measured in terms of the number of theoretical plates (N) in the column.

The combined effect of solvent efficiency and column efficiency is expressed in the *resolution* R_S of the column:

$$R_S = \frac{2(t_{R_2}-t_{R_1})}{w_1+w_2} \tag{2.54}$$

where t_{R_1} and t_{R_2} are the uncorrected retention times of components 1 and 2, and w_1 and w_2 are their basal peak widths (see Fig. 2.1). If the basal peak widths are approximately the same, Equation (2.54) reduces to:

$$R_S = \Delta t_R/w \tag{2.55}$$

In considering the effect of column length on resolution, it can be shown that:

$$\Delta t_R \propto \text{distance of migration down column} \tag{2.56}$$

and

$$w \propto \sqrt{(\text{distance of migration down column})} \tag{2.57}$$

so that Δt_R increases faster than w, and from Equation (2.54) we see therefore that the resolution can always be increased by an increase in column length. However, this will result in an increase in analysis time, unless the mobile phase velocity is increased to compensate for the increased column length.

The degree of resolution required will be determined to some extent by the nature of the chromatographic analysis to be performed. The simplest case is where two components are of more or less equal concentration. Reference to

Fig. 2.9 enables us to calculate that in this idealized situation, where the two peaks are triangular in shape, an $R_S = 1$ value would give complete separation at the base line. In practice, because the peaks are Gaussian in shape, an $R_S = 1.5$ value is taken to represent base line separation. Under the condition of equal concentrations of components 1 and 2, each component could be completely recovered with 99.9% purity. If the ratio of the concentrations of the two components was 10 to 1, component 1 could still be completely recovered in 100% purity, and component 2 could be 99% recovered in 99.7% purity. If the R_S value is reduced to $R_S = 1$, the cross contamination is only of the order of 3% and recovery rates are 98% for a 1 : 1 component ratio, falling to 88% for the minor constituent of a 10 : 1 component ratio.

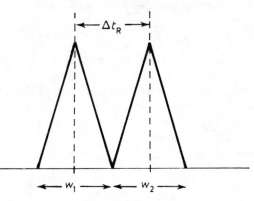

Fig. 2.9 Calculation of resolution.

An R_S value of 0.8 is normally considered the lowest practical value for qualitative analysis. Even at this value, for a 1 : 1 component ratio, 96% recovery at 95% purity can be achieved.

A chromatogram usually consists of more than two components, perhaps as many as fifty. It is therefore incorrect to refer to the resolution of a column in general terms. However, in practice, the resolution is quoted for the two components that are least well separated; the resolution for all other pairs of components is then at least as good as, if not better than, the quoted resolution.

The relationship between solvent efficiency as represented by the separation of the band centres, and column efficiency as represented by the band widths, is illustrated in Fig. 2.10. Fig. 2.10(a) represents the 'ideal situation' with both good solvent efficiency and good column efficiency giving a large separation of the two components and narrow band widths. In Fig. 2.10 (b), although the separation is the same, the band width is large and the resolution has been destroyed. Fig. 2.10 (c) shows that, even with poor column efficiency, complete resolution can be achieved as long as the solvent efficiency is sufficiently high.

Equation (2.54) is useful in cataloguing the degree of resolution but is purely an empirical relationship and does not show how resolution is related to the

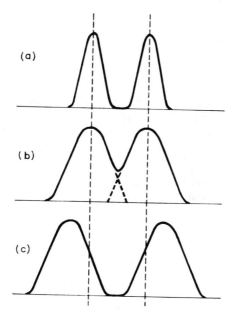

Fig. 2.10 The relationship between column efficiency and
solvent efficiency.

conditions of separation and cannot be used directly to improve resolution.
Purnell [10] has shown that for column chromatography the resolution (R_S) can
be related to the capacity factor (k'), the relative retention (α), and the number
of theoretical plates (N) in terms of the second component (subscript 2) of the
pair by the relationship:

$$R_S = \frac{1}{4}\left(\frac{\alpha - 1}{\alpha}\right)\left(\frac{k_2'}{1 + k_2'}\right)(N_2)^{\frac{1}{2}} \tag{2.58}$$

The resolution can therefore be changed by altering the thermodynamic
properties of the system (which alters α and k') or by altering the column
conditions (flow rate, particle size, etc.) to change N.

In order to investigate further the effect that these terms have on the resolution
it is useful to consider them as independent functions, thus:

selectivity $\qquad (\alpha-1)/\alpha$

capacity factor $\quad k_2'/(1+k_2')$

efficiency $\qquad (N_2)^{\frac{1}{2}}$

2.10.1 Effect of Selectivity on R_S

Equation (2.58) may be written:

$$R_S = c_1 \left(\frac{\alpha - 1}{\alpha} \right)$$

where c_1 represents constant capacity factor and efficiency and the variation of R_S with α may be obtained (Fig. 2.11). When $\alpha = 1$ there will be no thermodynamic difference between the two components ($K_1 = K_2$) and hence no separation. However, at least partial resolution has been achieved in systems with α as low as 1.01.

The main factor governing α is the stationary phase. Other factors are the mobile phase composition and, to a lesser extent, the temperature. Although

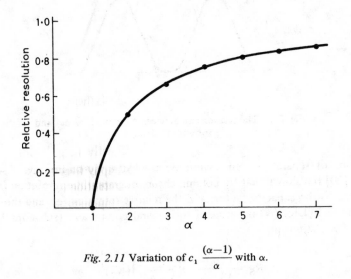

Fig. 2.11 Variation of $c_1 \dfrac{(\alpha - 1)}{\alpha}$ with α.

changes in resolution by selectivity changes are desirable, since they can often be achieved without an increase in analysis time or of column inlet pressure in a complex analysis, changes in α may merely lead to a reshuffling of peaks with no real improvement in the overall resolution.

2.10.2 Effect of Capacity Factor on R_S

Writing Equation (2.58) again in the form:

$$R_S = c_2 \left(\frac{k_2'}{1 + k_2'} \right)$$

where selectivity and efficiency are now constants, the variation of R_S with k' is represented in Fig. 2.12. Again there will be no separation if $k_2' = 0$. Increases

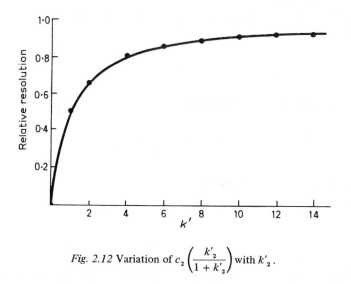

Fig. 2.12 Variation of $c_2 \left(\dfrac{k'_2}{1 + k'_2} \right)$ with k'_2.

in small k' values are seen to increase R_S very markedly, but at larger k' values the term $k'_2/(1+k'_2) \rightarrow 1$, and the term plays no further part in the resolution. Equation (2.20) shows that large k' values also mean long retention times and hence diffuse bands which are difficult to detect. The optimum range of k' values is therefore from 1 to 10. k' is mainly controlled by the mobile phase composition and hence gradient elution or solvent programming is used in liquid chromatography to achieve the same objective as temperature programming in gas chromatography.

2.10.3 Effect of Efficiency on R_S

Writing $R_S = c_3 . N^{\frac{1}{2}}$ the resolution increases as N increases. Note, however, that, since R_S is proportional to $N^{\frac{1}{2}}$ and N is proportional to the column length, a twofold increase in the column length will only increase R_S by a factor of 1.4, but the analysis time will be doubled, unless the inlet pressure is increased to give constant retention. Hence, column length is only used to increase N and R_S when other methods have failed. In practice it is usual to optimize the conditions for high N (Section 2.5) since this will always lead to improved resolution. N may also be increased by using a lower mobile phase flow rate (Fig. 2.7) but again the penalty is longer analysis times.

2.10.4   Effective Plate Number N_{eff}.

The number of theoretical plates N is not the best measure of the column efficiency since in measuring t_R the retention time (t_m) of an unretained peak has been included and this is a function of both the column and the extra-column configuration. A better measure of column efficency is given by the *effective plate number*, N_{eff}, where:

$$N_{eff} = 16\left(\frac{t_R - t_m}{w}\right)^2 \qquad (2.59)$$

N_{eff} is related to N by the relationship:

$$N_{eff} = \left(\frac{k'}{1 + k'}\right)^2 N \qquad (2.60)$$

and thus N_{eff} is always less than N.

Using N_{eff} instead of N Equation (2.58) then becomes:

$$R_S = \frac{1}{4}\left(\frac{\alpha - 1}{\alpha}\right)(N_{eff})^{\frac{1}{2}} \qquad (2.61)$$

and for a constant α, R_S is proportional to N_{eff}. Effective plate number is therefore a more useful parameter than N for the comparison of the resolving powers of different columns containing the same stationary phase.

Fig. 2.13 shows how N_{eff} varies with α. Where $\alpha \approx 1$, N_{eff} must be extremely large to provide resolution, but as α increases so N_{eff} falls very rapidly.

2.10.5 Effect of temperature on resolution

The variation of the equilibrium constant (K) with temperature (T) is given by the van't Hoff equation

$$\frac{d \ln K}{dT} = \frac{\Delta H}{RT^2} \qquad (2.62)$$

where ΔH is the enthalpy of solution from the mobile phase to the stationary phase. If the phase ratio V_s/V_m is independent of temperature, we may write the capacity factor k' for K. Hence:

$$\frac{d \ln k'}{dT} = \frac{\Delta H}{RT^2} \qquad (2.63)$$

Values of ΔH will be much smaller in liquid than in gas chromatography and hence temperature will only have a small effect on retention and resolution in liquid chromatography. Most analysis by liquid chromatography is therefore carried out at ambient or relatively low temperatures.

Temperature control is, however, useful in keeping k' values constant, in improving mass transfer by reducing the viscosity of the mobile phase and to a lesser extent in increasing the solubility of the mobile phase of sparingly soluble substances. However, since k' is strongly dependent on the nature of the mobile

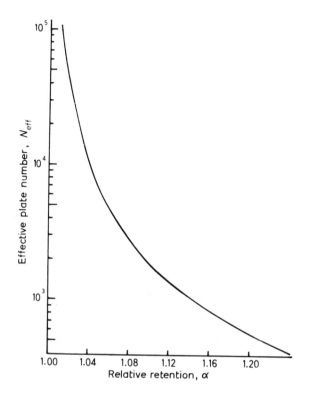

Fig. 2.13 Variation of effective plate number (N_{eff}) with relative retention (α).

phase, gradient elution, rather than temperature programming, is used to improve resolution in liquid chromatography.

2.11 Resolution and Time for Analysis

The time for an analysis is not necessarily an important criterion. However, since resolution and analysis time are interrelated, plates per second (N/t) or effective plates per second (N_{eff}/t) are sometimes better criteria for comparing column performance.

The time for an analysis is approximately equal to the retention time of the last component. The retention time can be written:

$$t_R = N(1 + k')(H/v) \qquad (2.64)$$

Fig. 2.14 shows a plot of H/v versus v to illustrate the effect of mobile phase velocity (v) on retention time. It can be seen that initially there is a sharp decrease in H/v as v increases, followed by a levelling off. Owing to the coupling of the A and C_M terms, however, H/v will continue to decrease as v increases. It could be assumed therefore that it was always advantageous to operate at as high a flow rate as possible, but in practice the column is normally operated at the 'optimum flow rate'. Unlike GC where v_{opt} is normally considered to be twice

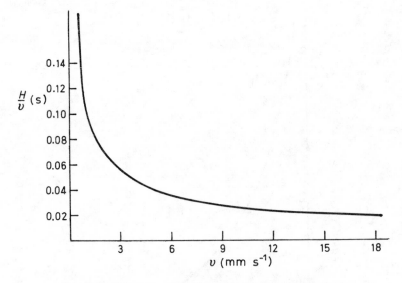

Fig. 2.14 Effect of mobile phase velocity on retention time.

the value of v_{min}, in LC this optimum value is not clearly defined, but it is usually taken to be the mobile phase velocity at which the H/v versus v curve shows a marked change in slope; this occurs when v is in the region $2-5$ cm s^{-1}.

The use of higher flow rates is usually limited by practical considerations. The flow rate is related to the pressure drop through the column (ΔP), column length (L), mobile phase viscosity (η), and column permeability (K°), by the expression:

$$v = \Delta P K^\circ / \eta L \qquad (2.65)$$

In LC mobile phase viscosities are some 100 times greater than in GC, so pressure drops in LC will be 100 times greater than in GC for a given column length and flow rate. This mainly accounts for the use of high pressures in liquid compared with gas chromatography.

The Kozeny–Carman equation [11, 12] gives the column permeability in a regular packed column (one in which the ratio of column diameter to particle diameter is greater than 10):

$$K^\circ = \frac{d_p{}^2}{180} \cdot \frac{\epsilon^3}{(1-\epsilon)^2} \qquad (2.66)$$

where ϵ is the interparticle porosity. For a packed column $\epsilon = 0.42$, so Equation (2.66) can be written [13]:

$$K^\circ \approx d_p{}^2 / 1000 \qquad (2.67)$$

In GC the typical particle diameter is $100-150 \, \mu\text{m}$, whereas in LC, particle diameters of $5-10 \, \mu\text{m}$ are in common use. Hence, permeabilities are of the order of one hundred times lower in liquid chromatography.

The combined effect of viscosity and particle size means that inlet pressures of $500-5000 \, \text{lbf in}^{-2}$ would be required if short columns were not used.

By substituting for v in the retention Equation (2.20), from Equation (2.65) we have:

$$t_R = \frac{\eta L^2 (1 + k')}{K^\circ \Delta P} \tag{2.68}$$

or from Equation (2.62)

$$t_R = \frac{1000 \eta (1 + k')}{\Delta P} \left(\frac{L}{d_p}\right)^2 \tag{2.69}$$

Therefore, for constant values of L, k', η, and ΔP, shorter retention times are obtained with more permeable columns. Alternatively, if t_R is kept constant a more permeable column will permit a longer column to be used, with a consequent increase in N and better resolution. For a regular packed column the permeability can be increased by increasing the particle diameter, but larger particles mean higher H values. Halasz and Walkling [14] and Knox and Parcher [15] have shown that when $d_c/d_p < 5$, although the permeability increases some tenfold, the H values actually decrease owing to increased radial mixing in the mobile phase.

Combining Equations (2.60) and (2.64) an expression is obtained for the number of effective plates per second:

$$\frac{N_{eff}}{t} = \frac{v}{H} \cdot \frac{(k')^2}{(1 + k')^3} \tag{2.70}$$

A plot of $(k')^2/(1 + k')^3$ versus k' is given in Fig. 2.15.

Assuming that H is independent of k' (as it will be when mass transfer in the mobile phase is the predominant band broadening factor), the k' function is directly related to N_{eff}/t. The optimum k' value is then seen to be about 2, but there is little loss in speed of analysis for higher k' values. However, k' values less than about 1.5 are to be avoided if N_{eff}/t is not to suffer.

2.12 Theory of Exclusion Chromatography

Exclusion chromatography separates molecules according to size. The gel particles consist of a porous matrix with a closely controlled pore size. The channels between the gel particles are much larger than the pores of the gel itself, so large molecules whilst passing freely through the column cannot enter the pore structure of the gel. Smaller sample molecules and the solvent molecules may be free to penetrate the entire pore volume and will be retarded to the greatest extent. Hence

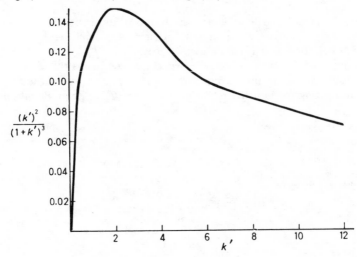

Fig. 2.15 Plot of $(k')^2/(1+k')^3$ versus k'.

the larger molecules are eluted first from the column followed by the smaller molecules, with the solvent last.

This process is illustrated in Fig. 2.16 which is a calibration curve for an exclusion separation. The point 'A' represents the exclusion limit of the packing. Molecules with a molecular weight larger than the value given by 'A' are totally excluded from the pores and elute as a single peak (C) with retention volume V_m (the interstitial, interparticle or void volume). Point 'B' represents the molecular weight at which total permeation of the packing occurs. Thus all molecules with a lower molecular weight than 'B' will again elute as a single peak (F) with a retention volume V_t (the total permeation volume) where $V_t = V_m + V_s$ and V_s is the volume of the stationary phase (i.e. the volume of mobile phase contained in the porous particles or intraparticle volume). Compounds with molecular weights between these two limits may be separated—peaks D and E. The range of molecular weights corresponding to a retention volume $A < V_r < B$ is the fractionation range of the gel.

In exclusion chromatography the distribution coefficient of a solute species is defined [16] in terms of the fraction of the intraparticle volume, V_s, which is accessible to the solute so that:

$$K_d = \frac{V_r - V_m}{V_s} \tag{2.71}$$

and

$$V_r = V_m + K_d V_s \tag{2.72}$$

For large, totally excluded molecules, $V_r = V_m$ and hence $K_d = 0$. For small solute molecules which can enter all the pores, $V_r = V_m + V_s$ and hence $K_d = 1$.

Separation therefore occurs only where solute molecules obey the condition

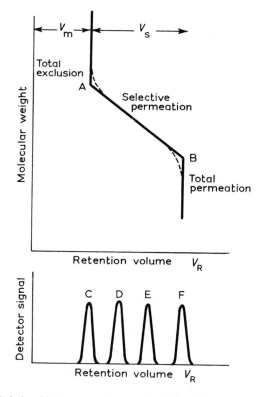

Fig. 2.16 Relationship between solute molecular weight and retention volume for 'ideal' exclusion chromatography.

$0 < K_d < 1$. The number of peaks resolvable is therefore strictly limited and has been estimated by Giddings [17] to be between 10 and 20.

The K_d value of a solute molecule is determined by its size; this depends on the molecular hydrodynamic radius i.e. the radius of gyration of the molecule in the solvent. Hence the effective size depends on such factors as the geometric shape (e.g. spherical, coiled or rod-like), solute–solute association and solvation.

As a general rule it is found that unless there is a molecular weight difference between compounds of at least 10% no resolution will be achieved.

K_d is independent of column size and geometry within a given gel–solvent system, but anything that would change the pore size of the gel, i.e. the nature of the gel, the solvent, and the temperature, would change the K_d of the solute. If the value of $K_d > 1$ then some other mechanism, other than exclusion, is contributing to the retention. Generally, adsorption is the explanation, though partitioning may occur if mixed solvents are used or if the solvent is chemically very different from the gel.

In the absence of adsorption effects exclusion chromatography gives Gaussian-shaped peaks and all peaks must be eluted within the total permeation volume, V_t. This makes it very easy to use automatic injection and fraction collecting.

The efficiency of a gel permeation column is markedly influenced by the mobile phase flow rate as shown in Fig. 2.17. In contrast to the other LC modes,

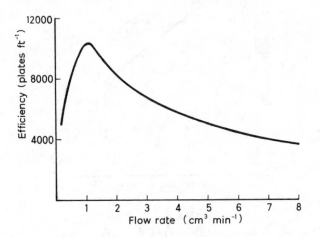

Fig. 2.17 Variation of efficiency with flow rate in exclusion chromatography with 10 μm Styragel.

the efficiency versus flow rate curve shows a maximum; and there are obvious disadvantages in working at high flow rates.

The advantages of exclusion chromatography may be listed as follows.
(1) Retention times are predictable according to molecular size.
(2) Band widths are narrow and detection is therefore facilitated.
(3) Analysis times are short and gradient elution is not required.
(4) Because of the inert nature of the gels, sample loss or reaction on the column does not occur.
(5) Column life is long.

References
1. Giddings, J. C. (1965) *Dynamics of Chromatography*, Marcel Dekker, New York.
2. Halasz, I. (1971) in *Modern Practice of Liquid Chromatography* (ed. J. J. Kirkland), Wiley-Interscience, New York.
3. Huber, J. F. K. (1969) *J. Chromatog. Sci.*, 7, 85.
4. Halasz, I., Endele, R. and Asshauer, J. (1975) *J. Chromatog.*, 112, 37.
5. Grushka, E., Snyder, L. R. and Knox, J. H. (1975) *J. Chromatog. Sci.*, 13, 25.
6. Snyder, L. R. (1972) *J. Chromatog. Sci.*, 10, 369.
7. Done, J. H., Kennedy, G. J. and Knox, J. H. (1972) in *Gas Chromatography* (ed. S. G. Perry), Applied Science Publishers.
8. Deininger, G. and Halasz, I. (1970) in *Advances in Chromatography* (ed. A. Zlatkis), Preston Technical Abstracts Co., Chicago.

9. Scott, R. P. W. and Kucera, P. (1971) *J. Chromatog. Sci.*, **9**, 641.
10. Purnell, J. H. (1960) *J. Chem. Soc.*, 1268.
11. Kozeny, J. S. B. (1927) *Akad. Wiss. Wien*, Abt. IIa **136**, 271.
12. Carman, P. C. (1937) *Trans. Inst. Chem. Eng.* (London), **15**, 150.
13. Halasz, I. and Heine, E. (1968) in *Progress in Gas Chromatography* (ed. J. H. Purnell), Interscience, New York.
14. Halasz, I. and Walkling, P. (1969) *J. Chromatog. Sci.*, **7**, 129.
15. Knox, J. H. and Parcher, J. F. (1969) *Anal. Chem.*, **41**, 1599.
16. Wheaton, R. M. and Bauman, W. C. (1953) *Ann. New York Acad. Sci.*, **57**, 159.
17. Giddings, J. C. (1967) *Anal. Chem.*, **39**, 1027.

3

Equipment

3.1 Introduction

Classical liquid chromatography uses simple apparatus: a solvent reservoir, a wide bore glass column, a relatively large diameter particle packing, and collection tubes for collecting the eluent. Detection is usually by chemical and physical methods and large samples are used. To increase the speed of analysis and resolving power to match that of GC, it is necessary to increase the low liquid flow rates used and to obtain faster equilibration of the sample between the mobile and stationary phases so that resolution keeps pace with the faster flow rates. These two aims are achieved by pressurizing the mobile phase and by using packings of small particle size. Using smaller particle size packings decreases the permeability of the columns, so that to avoid excessively high pressures, shorter columns must be used. The diameter of the columns may also be reduced as this gives the maximum linear flow velocity for a given volume flow rate. Because of the smaller amounts of stationary phase in the column, sample sizes must be reduced to prevent column overloading. A major breakthrough has been the development of sensitive on-line detection systems capable of monitoring micro-quantities resolved from very small samples using small capacity ($<10 \ \mu l$) flow cells to prevent loss of resolution.

The modern liquid chromatograph then consists basically of a high pressure pumping system, relatively narrow bore short columns packed with a small particle size stationary phase, and an on-line highly sensitive detector system. A general schematic of a typical LC chromatograph is shown in Fig. 3.1.

3.2 Mobile Phase (Solvent) Reservoirs and Solvent Degassing

For analytical separations the solvent reservoir should be of about 1 dm^3 capacity. For preparative applications a much larger reservoir may be required. The reservoir may be constructed from any material; glass is cheapest but is

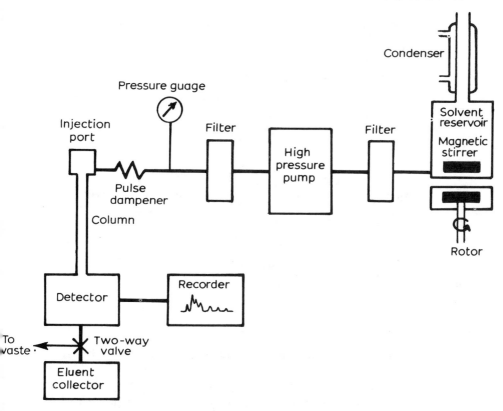

Fig. 3.1 Schematic diagram of a typical high performance liquid chromatograph.

subject to breakage, and stainless steel may be used. Before the solvent is used it should be degassed. Degassing is required to remove dissolved gases (in particular oxygen) which may react with either the mobile or the stationary phase. It is also necessary to prevent degassing in the detector, where the solvent pressure may reduce to atmospheric, to prevent base-line drift or continuous spikes. These effects are particularly noticeable at the lower region of the ultra-violet (below 220 nm).

The solvent is usually degassed *in situ*. The degassing system may consist simply of a hot-plate and magnetic stirrer placed underneath the glass reservoir to which is fitted a reflux condenser. A more elaborate commercial unit is shown in Fig. 3.2. Degassing is accomplished by means of a built-in vacuum pump. The pressure over the solvent can be regulated by a bleed valve, and the solvent temperature is controlled by an immersion heater and a resistance thermometer. The reflux condenser prevents loss of solvent and also avoids composition changes during treatment of mixed solvent systems. To prevent the re-adsorption of air or water from the atmosphere the system can be purged with a dry purge gas.

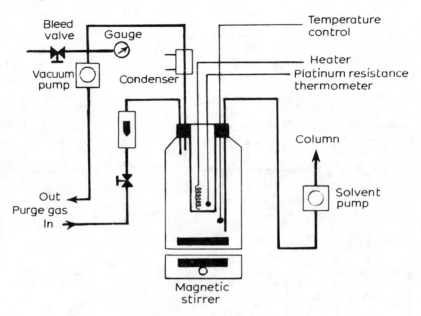

Fig. 3.2 Solvent storage and degassing system (after Hewlett Packard).

Other degassing units use ultrasonic vibrations to remove the dissolved gases whilst others use a helium sparge.

Although solvent refluxing has been found to be the most efficient way of removing oxygen it is inconvenient and potentially hazardous. The use of a helium sparge to displace dissolved gases is increasing, but its effectiveness varies with the nature of the mobile phase and it is necessary to continue sparging to prevent re-adsorption. Vacuum degassing too gives variable results with different mobile phases. Ultrasonic degassing although cheap and convenient is not very effective and also needs a back-up purge to prevent re-adsorption.

3.3 Pumping Systems

The development of suitable pumping systems has been one of the main factors in the development of modern liquid chromatography, and the pump is a critically important component. There are two basic types of pump in common use: constant pressure pumps and constant volume pumps. The requirements for a pumping system are listed in Table 3.1. Neither type of pump however satisfies all the listed criteria.

3.3.1 Constant Volume (Constant Displacement) Pumps

If a constant volume pump is used, changes in the permeability of the system, caused by settling or swelling of the packing, or viscosity changes in the mobile phase (due to temperature fluctuations or composition changes) are compensated for by pressure changes and the flow rate remains constant. Because flow changes

Table 3.1   *Requirements for Modern LC Pump*

1. Maximum pressure range about 6000 lbf in^{-2}
2. Pulse-free or fitted with a pulse dampener.
3. Flow delivery of between 1 and 10 cm^3 min^{-1} for analytical applications and up to 200 cm^3 min^{-1} for preparative applications.
4. Solvent flow control to ±0.5%.
5. Flow rate reproducibility better than 0.5%.
6. Chemically inert to common solvents; i.e. stainless steel and Teflon seals must be used.
7. Small volume (~0.5 cm^3 maximum) pumping chamber for gradient elution and re-cycle devices.

cause non-reproducible retention times, adversely affect resolution, and give unstable base-lines, the constant volume pump provides a more precise analysis. It is particularly useful when gradient elution is used. There are two main types of constant volume pumps: single stroke (syringe type) pumps, and reciprocating pumps having either a diaphragm or a piston.

The syringe type pump consists of a syringe the plunger of which is driven by a stepping motor through a gear box. The rate of delivery from the syringe is controlled by varying the voltage on the motor. Fig. 3.3 is a schematic diagram of a single stroke syringe pump. The main advantage of this type of pump is that it is capable of providing a pulse-free flow at high pressure (up to 7500 lbf in^{-2})

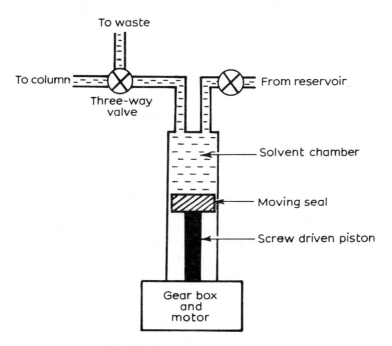

Fig. 3.3 Single stroke syringe pump.

and the flow rate is independent of the operating pressure, if the compressibility of the liquid ($\sim$3% at 6000 lbf in^{-2}) is ignored. Its main disadvantage is that it has a finite solvent capacity. However, since this is typically $\sim$500 cm^3 and a typical chromatogram requires 20–40 cm^3, this is not a great disadvantage. Dual syringe systems found in some instruments are of course expensive, but with suitable gradient formers they also provide the gradient elution capacity. Due to its high cost and lack of flexibility this type of pump is little used in today's commercial instruments.

Reciprocating pumps fall into two types: diaphragm and piston pumps. Fig. 3.4 is a schematic diagram of an Orlita diaphragm pump. A working piston and a

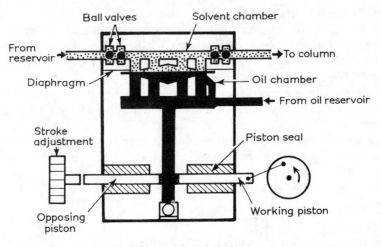

Fig. 3.4 Orlita dual piston diaphragm pump.

floating piston oppose each other in an oil chamber which is separated from the solvent chamber by a stainless-steel diaphragm. The delivery rate of the pump is adjusted by limiting the travel of the opposing piston to some fraction of the working piston. The solvent is in contact with only the stainless-steel diaphragm. Check valves at the outlet and inlet of the solvent chamber are an essential part of the pump design and must function well if the pump is to deliver an accurate flow of solvent. The diaphragm pump avoids one of the main problems with the piston pumps—wear of the piston seal and the subsequent deposition of seal material in the check valves which prevents them from functioning properly.

Fig. 3.5 is a schematic diagram of a piston type constant volume pump. Each forward stroke of the piston pushes solvent through the non-return valve on the column side, and on the reverse stroke more solvent is drawn in from the solvent reservoir.

The main advantages of reciprocating pumps is that their internal volume can be made very small and their delivery is continuous. The flow rate is constant regardless of the back pressure from the LC column but deviations from constant

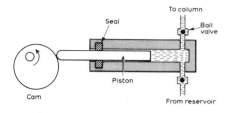

Fig. 3.5 Piston type constant volume pump.

flow are often observed at high column pressures.

The main disadvantage of reciprocating pumps is that the solvent delivery is not pulse free, and this shows up as base-line noise, particularly when a refractive index detector (refractometer) is being used. Pulsing can be alleviated by the use of two or more pump heads operating out of phase but feeding into a common supply, or by the use of pulse dampeners.

3.3.2   Constant Pressure Pumps

Constant pressure pumps, consisting of some form of pneumatic device for the direct pressurization of the mobile phase with an inert gas, give a reliable pulse-free flow and have the advantage of low cost and simplicity. They are however not as accurate as constant volume pumps but can be used where flow accuracy and reproducibility are less critical.

The simplest type of constant pressure pump is the pressurized coil pump or gas displacement pump shown in Fig. 3.6. The hydraulic chamber consists of a stainless steel coil. The solvent volume can therefore be varied by using coils of different volume although the normal volume of coil used $(200-500 \mathrm{~cm}^{-1})$ is sufficient for most analyses. The coil can be gravity filled from an external reservoir through the three-way valve B, and in operation the solvent in the coil is displaced by nitrogen at high pressure. The gas flow can be made to by-pass the solvent coil by switching the two-way valve A. During use, the displacing gas will dissolve in the solvent at the liquid—gas interface, but this portion of the solvent can be removed through tap C at the next filling. The inlet pressure is limited by the available gas pressure (usually up to $2500 \mathrm{~lbf~in}^{-2}$).

The pneumatic amplifier pump (Fig. 3.7) is a more elaborate form of pneumatic pump. It gives a pulse-free flow during its displacement stroke, can be filled rapidly, and can deliver solvent at high flow rates. It is not limited by gas pressure in the way that the coil pump is limited. Pressure from an external source is supplied to a large surface area gas piston attached to a smaller surface area hydraulic piston. Thus the applied pressure is amplified and a column pressure of $5000 \mathrm{~lbf~in}^{-2}$ can easily be obtained with a $100 \mathrm{~lbf~in}^{-2}$ gas supply. The gas pressure is released at the end of the delivery stroke and the solvent vessel refills by the action of gas pressure or a spring. Check valves between the solvent chamber and the solvent reservoir, and between the solvent chamber and the

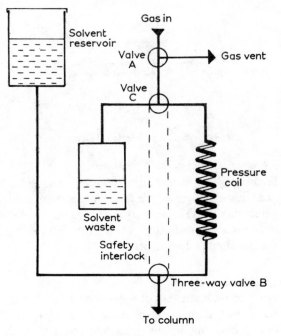

Fig. 3.6 Pressurized coil pump.

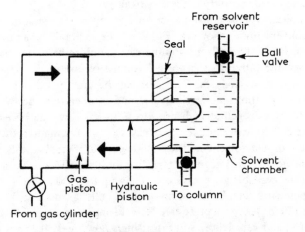

Fig. 3.7 Pneumatic amplifier pump.

column, prevent back flow during re-filling. The delivered flow from a constant pressure pump is very sensitive to changes in column permeability and to changes in solvent viscosity. This makes them unsuitable for gradient elution work where the mobile phase viscosity may vary considerably, unless some form of automatic flow control is incorporated into the system.

3.4 Flow Controllers

Closed loop (feed-back) flow controllers are available on a number of commercial instruments. They consist of a flow-through pressure transducer, which measures flow rate by measuring the pressure drop across a restrictor of fixed value, placed at the pump outlet. The flow rate signal is fed back to a control unit which compares the actual and the pre-set flow rate. Any difference is countered by a change in pump motor speed, or a change in gas pressure on the pneumatic piston, according to the type of pump. Thus, changes in mobile phase flow rate caused by column permeability changes, viscosity changes, or temperature changes are corrected. When coupled with a gradient elution accessory, precise flow control can be achieved regardless of viscosity changes resulting from composition changes during the gradient elution. The response time of the controller is less than 1 second and is much faster than the pump noise caused by reciprocating pumps, and hence flow pulsations caused by reciprocating pistons or diaphragms are eliminated.

3.5 Solvent Flow Programming Equipment

In gas chromatography the general elution problem is often overcome by the technique of temperature programming. In liquid chromatography, the technique does little to improve the separation relative to isocratic elution. Instead, some property of the solvent is programmed. This may be brought about by changing the flow rate of the solvent during the separation (flow programming) or by changing the strength (polarity, pH, or ionic strength) of the solvent (gradient elution or solvent programming). Many commercial solvent programmers can control both of these techniques.

Flow programming. Flow programming requires control of the output of a single solvent pump. The way that this is done depends on the type of pump used. With a constant displacement (syringe type) pump the voltage to the stepping motor can be varied at a predetermined rate. On a dual-piston reciprocating pump, the delivery rate can be varied by an electro-mechanical drive system.

Gradient elution (Solvent programming). Gradient formers are conveniently divided into two types according to whether the mixing of the solvents takes place at low (atmospheric) pressure or at high pressure.

3.5.1 Low Pressure Gradient Formers

Low pressure gradient formers have the advantage of simplicity and cheapness. Only one high pressure pump is required to transport the solvent from the mixing chamber to the column. A simple three-solvent programmer is shown schematically in Fig. 3.8. Vessel A would contain the solvent of lowest solvent strength and vessel C that with the greatest solvent strength. The strength of the solvent is increased by sequentially opening the connecting valves AB and BC and allowing the solvents to mix. By increasing the number of solvent reservoirs more complex gradients can be formed.

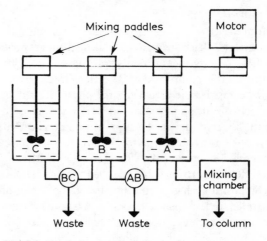

Fig. 3.8 Simple low pressure three-solvent programmer.

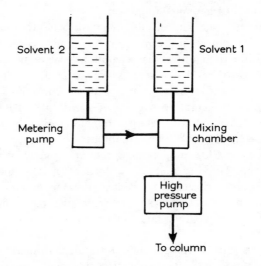

Fig. 3.9 Gradient elution with a single high pressure pump.

A second system is shown schematically in Fig. 3.9. The low strength solvent (1) is connected directly through the mixing chamber to the high pressure pump and the column and can be used directly for isocratic elution. When a gradient is required the higher strength solvent (2) is fed into the mixing chamber through the metering pump. Additional solvent reservoirs with suitable switching valves can be added as required.

Low pressure mixing with time proportioning valves is now used on several commercial instruments. This requires very accurate value timing and fast operation and the success of this type of system has been achieved with the use of microprocessor control. However, the overall effectiveness of the gradient system depends on the efficient mixing of the liquids, since a slug of liquid A is followed by a slug of liquid B (and even by C in a ternary gradient system). Often two low volume stirred mixing chambers are used in series for this purpose.

3.5.2 High Pressure Gradient Formers

In this approach the separate solvents are pumped, at high pressure, into a mixing chamber connected to the inlet side of the column. The advantages of this system are that any type of gradient can be generated simply by programming the delivery of each pump, and it can be used effectively to determine the optimum solvent composition for an isocratic elution. The method has the disadvantage of being more expensive, since it usually requires two high pressure pumps, and is normally limited to gradients that can be achieved by mixing two solvents. A typical schematic diagram is shown in Fig. 3.10. This system also uses a feed-back flow controller to maintain constant

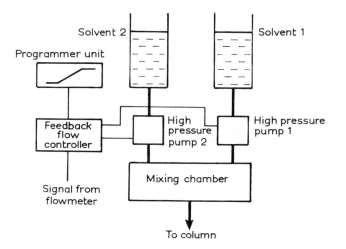

Fig. 3.10 Gradient elution with two high pressure pumps.

flow. The gradient elution programmer unit can be designed to operate either with two constant pressure pumps or with two constant volume pumps. When constant pressure pumps are used, solenoid valves allow for the flow to be alternatively switched on and off. With constant volume pumps interlaced pulses can be provided to the two pumps and the total flow maintained constant.

Gradient systems using syringe pumps can be subject to inaccuracies and lack of reproducibility due to the compressibility of the solvents. The syringe pumps control the speed of the piston and not the flow of solvent from the pump. If the back pressure on the column changes owing to viscosity changes in the mobile phase, the solvent in a large volume syringe pump can undergo considerable compression or expansion. This can lead to deviations of the actual form of the gradient from the required form. Flow rate changes with solvent viscosity may also be observed. These problems can be reduced by always using the same initial volume of solvents in the two syringes.

Reciprocating pumps, since they do not have a high volume of solvent under high pressure, do not suffer the same compressibility problems. However, they do suffer from problems due to their inability to pump very small volumes (<0.04 cm^3 min^{-1}) and to flow inaccuracies at very low and very high flow rates.

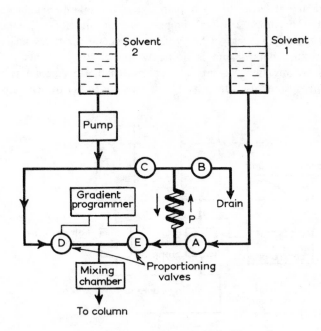

Fig. 3.11 Gradient former with holding coil.

A third gradient former system is shown in Fig. 3.11. Solvent 1 is held in a holding coil which is filled from an external reservoir through taps A and B. Solvent 2 is pressurized by a single high pressure pump which transports solvent to the mixing chamber through tap D. With tap C open, solvent 2 also displaces solvent 1 from the holding coil through tap E to the mixing reservoir. The gradient controller operates the on–off proportioning valves D and E which determine the mixing ratio in the mixing chamber. As long as the volume of the holding coil P is not too great this system is not subject to large compressibility effects, but its application is limited in automatic operation and in preparative work by the volume of the holding coil. The flow rate may also change as a result of viscosity changes in the solvent unless a feed-back flow controller is incorporated.

Because of slow diffusion in liquids, and the relatively low flow velocities, turbulent mixing of the solvents is rarely achieved. To achieve rapid and efficient mixing of the solvents therefore the volume of the mixing chamber must be small.

3.6 Pulse Damping

A pulsating mobile phase can cause unwanted disturbance to certain detectors (e.g. differential refractometers) and it can limit detector sensitivity according to the amount of base-line noise it causes.

Pulse damping is less of a problem with multi-headed reciprocating pumps since their combined delivery is inherently more constant, particularly if sinusoidal drive cams are used to keep flow rates constant, but the initial cost is considerably higher.

Pulse dampers can either be in-line or out-of-line. A simple in-line pulse damper may be made from a length of stainless-steel capillary tubing (10 metres X 0.25 mm (0.01 inch) i.d.) placed after the pump. Out-of-line dampers range from a simple air-filled snubber made from a 50 cm length of 5 mm bore stainless-steel tubing closed at one end through an adjustable spring-loaded hydraulic snubber, to the hydraulic analogue of an electrical smoothing choke. This latter can be made from a series of Bourdon tubes. A combination of in-line restrictor and Bourdon tubes may be used to great effect. The restrictor can be made from a length of stainless-steel capillary tubing (5 metre X 0.25 mm i.d.), and the Bourdon gauge used for pressure measurement provides the Bourdon tube. A series of three such combinations of restrictor and Bourdon tubes can reduce the noise to less than 1% of full scale deflection and allow the operation of both u.v. and RI detectors at full sensitivity over a wide pressure range.

The volume of the pulse damping system increases the inconvenience of solvent change-over because the extra volume must be flushed with the new solvent. A more serious problem may occur during gradient elution when the extra volume acts as a mixing vessel which may affect the decreed solvent gradient.

3.7 Pressure Measurement

The column inlet pressure should be monitored by a suitable device such as a Bourdon gauge or a pressure transducer. The Bourdon gauge is robust, cheap, and simple but is less precise than the pressure transducer. When used with constant volume pumps the gauge should be protected by a pressure cut-out on the pumping system to protect it from excessive pressures. The dead volume of the Bourdon gauge can also be troublesome, both after changing solvent, when small amounts of residual solvent can cause base-line drift for a considerable time, and as a source of air which, even after solvent degassing, may cause trouble with the refractive index detector. Pressure transducers are more precise, have a low dead volume, and because of their flow-through nature, cause fewer problems on solvent change-over. They can also be provided with a pressure cut-off to protect the pumping system. They are however rather more expensive.

Pressure monitoring is more important in liquid chromatography than in gas chromatography. Changes in column inlet pressure are indicative of plugging of the column, of the pipe-lines, or of the detector, or of a malfunction of the pumping system. A knowledge of the pressure is also necessary for calculating the permeability of the column and as an aid to the optimization of column parameters.

3.8 Filters

Filtration of all solvents used in HPLC is essential. Not only does particulate matter plug the very narrow bore tubing used, particularly between a loop injector and the column, but it also reduces the life of pump seals and causes valve problems. To avoid flow restriction from the solvent reservoir to the pump a large area 2 μm stainless-steel filter is used in the solvent reservoir.

Debris from pump seals also needs to be filtered out so that it does not accumulate on the top of the column. A small area 2 μm frit is suitable here as some pressure drop is acceptable on the high pressure side. Pre-columns also act to filter out particulate matter as well as very polar contaminants in the mobile phase.

3.9 Sample Introduction Systems

For maximum efficiency on a chromatographic column the sample should be introduced, ideally, as an infinitely narrow band. This requirement is even more stringent in liquid chromatography than in gas chromatography; solutes with small k' values are particularly susceptible to extra-column band broadening.

3.9.1 Syringe Injectors

On-column syringe injection leads to the highest separation efficiency. If specially designed syringes are used, syringe injections can be made up to pressures of at least 1500 lbf in^{-2}; ordinary microlitre syringes as used in gas chromatography can often be used up to 1000 lbf in^{-2}. Syringe injections can be made through a septum injector (Fig. 3.12) or using a septum-less system. With a septum injector the internal volume should be as small as possible and should be swept completely by the mobile phase. The septum must be compatible with the mobile phase and to achieve this it is often necessary to use 'sandwich septa' consisting of a neoprene septum with Teflon on one face. Fluoroelastomeric and silicone septa can also be used. The septum may be the pressure limiting factor but the use of dual septa and stainless steel discs pressing down on the septa (see Fig. 3.12) enables pressures up to 3000 lbf in^{-2} to be used.

When using an 'infinite diameter' type column the solute must be injected precisely into the centre of the column. The septum injector should therefore be modified to include a needle guide to achieve this. It will also ensure that the septum will be pierced repeatedly in the same spot thus increasing its life. When using on-column septum injection it is important to ensure that the needle does

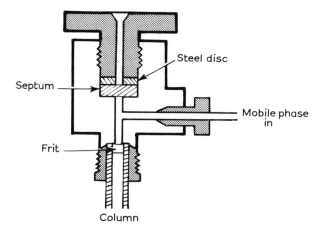

Fig. 3.12 Syringe injector.

not pierce into the column packing itself. Breaking up the column in this way may lead to distortion of the peak shape as well as blocking of the syringe needle. Small pieces of septum may also be deposited on the top of the column. The use of a rigid frit on the end of the column prevents this although it does lead to a small decrease in column efficiency.

Septum-less or stop-flow injection avoids many of the foregoing problems. This involves either switching off the pump, allowing the pressure to fall to atmospheric, injecting the sample, and then re-starting the pump, or using a three-way valve to divert the flow whilst the injection is made. However, the stop-flow technique can introduce retention time uncertainty and cause flow disturbances in some detector systems.

3.9.2 Valve-Loop Injectors

Valve injectors are designed to operate at pressures up to 7000 lbf in^{-2} without the use of septa. The solvent flow is by-passed into the column and an external or internal loop is filled with sample, which is then introduced into the column by switching a valve. The sample loops can either be external and interchangeable, to give a range of volumes (10 to 500 μl), or internal and of fixed volume. The fixed volume values are used primarily for very small samples (1–3 μl), the sample volume being engraved in the valve seal.

The advantages of the loop–valve injections are: (i) the ability to inject a wide range of sample volumes with a high degree of reproducibility; (ii) the capability of injecting at high pressures ($\sim$ 7000 lbf in^{-2}) without stopping the solvent flow; (iii) the absence of a septum to cause column plugging; and (iv) the possibility of automation for use with automatic sampling systems. Their main drawback is that a reasonable volume of solute is required; it takes about 100 μl to fill a 10 μl sample loop properly.

3.9.3 Syringe-Loop Injectors

Syringe—Loop injectors combine a loop injector with a microsyringe and zero volume fitting (Fig. 3.13). This enables the injector to be used either as a conventional fixed volume loop injector, or as a variable volume injector. The sample loop is first filled with mobile phase. The zero dead volume fitting allows the sample to be injected into the sample loop without any sample loss, displacing an equal volume of mobile phase out of the dip tube. The valve is then rotated and the sample flushed onto the column in the normal way. It is possible to inject samples from less than 1 μl up to half of the sample loop capacity without loss of precision and accuracy. With larger samples some loss of sample from the sample loop may occur. Since the syringe never comes into contact with high pressures there is no restriction on the operating pressure due to the use of a syringe.

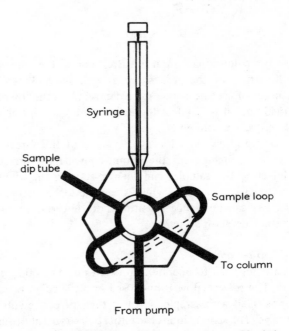

Syringe

Sample
dip tube

Sample loop

To column

From pump

Fig. 3.13 Syringe-loop injector (after Spectra Physics).

3.10 Columns and Column Fittings

The total amount of band broadening in the chromatograph is the sum of the band broadening of the various components and is expressed in terms of the variance (σ^2) of the peak width:

$$\sigma^2_{\text{total}} = \sigma^2_{\text{injector}} + \sigma^2_{\text{column}} + \sigma^2_{\text{connections}} + \sigma^2_{\text{detector}}$$

The peak width due to the equipment other than the column may be measured by connecting the detector and its connections directly to the injector. As discussed in Section 2.9, this extra-column band broadening places strict constraints on the injector, connecting tubing and detector volumes.

The optimization of columns and column fittings is particularly important when microparticle packings ($<10\,\mu$m diameter) are used.

3.10.1 Columns

The optimum diameter of the column will depend on the peak volume and how well the column is packed. Even with a well-packed column there will be a region near the wall–packing interface where solute band broadening will be increased. The extent of this region is reduced by using polished precision bore tubing and a special 316 stainless-steel tubing is available as a chromatographic quality. In order that 'wall effects' should not significantly affect column performance the minimum internal diameter should be about 5 mm. Most manufacturers approach this with a 4.6 mm column and columns of 2.1, 3.2 and 4.1 mm are also available for analytical use. When inlet pressure is limited it should be remembered that a 2.1 mm i.d. column will require about five times the inlet pressure as the same length of 4.6 mm i.d. column for the same volume flow rate (although the linear velocity will not be the same). As column diameter is increased the extra-column band broadening becomes relatively less important and it is easier to optimize the column performance.

For semi-preparative and preparative scale separations, columns up to 30.0 mm i.d. are used. However, the efficiency of these larger bore columns is reduced and for many purposes a smaller bore column can be used. The important parameter with preparative columns is the sample load. The loading capacity of a 4.6 mm i.d. column will be almost five times greater than a 2.1 mm i.d. column and a 9.4 mm i.d. column will have a loading capacity more than nine times greater than a 4.6 mm i.d. column.

Microbore columns with an internal diameter of 1.0 mm and up to 100 cm in length have recently been introduced. These are packed with normal 5 and 10 μm particle diameter and produce plate counts up to 25 000 per metre for 10 μm silica. One of their main advantages is that at the low flow rates used (10–100 μl min^{-1}) the consumption of mobile phase is very low which could be an advantage in liquid chromatography – mass spectrometry systems. The low mobile phase volume also reduces the dilution of the sample and detection limits may be improved. The main disadvantages are that analysis times are longer and the columns cannot be used in a standard chromatograph. Modifications can be made, but the full potential of these columns will not be realized until pumps and detectors fully compatible with the columns are available.

For normal analytical columns straight sections of columns between 10 cm and 100 cm in length are commonly used. Most manufacturers have standardized on a 25 cm column length; these columns are capable of generating up to 20 000 plates, however the use of shorter columns giving smaller pressure drops and shorter analysis times may in many analyses be sufficient (many published results require less than 10 000 plates). U-shaped and coiled columns result in a loss in efficiency but as long as the ratio of the radius of curvature of the coil to that of the column is greater than 130 the loss in efficiency is only small. The column should be packed before coiling. Larger columns are usually made up by joining several lengths of column together using connectors having zero dead volume. Since the overall efficiency of the combined column will be similar to that of the worst of the short columns, the columns should be matched for efficiency. Similarly columns of different diameters and different stationary phases should not be connected.

3.10.2 Column Fittings

The column fittings found to be the best are those that have a minimum dead volume and no pockets where the sample can collect in stagnant pools or be diluted by the mobile phase. In a 'zero dead volume' fitting both the connecting tube and the column touch the frit, thus eliminating all dead volume. Both 'male' and 'female' union fittings are available, but the female type (Fig. 3.14) has certain advantages.

1. The female connection type is less susceptible to thread damage when columns are reconnected several times.
2. The ferrule is pressed against a cone and is not subject to deformation, nor does it swage into the column wall and cause deformation.
3. The internal volume can be made very small.

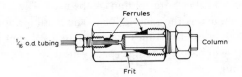

Fig. 3.14 Column connector with 'female' fittings.

The connecting tubing must be of small bore (0.01–0.02 inch i.d.) to prevent sample band spreading.

To retain the packing in the column, porous plugs or frits are used. The outlet closer should be a stainless-steel frit or gauze with a mesh size of 2–5 μm depending on the packing diameter.

If the column is to be used only with a valve or stop-flow injector the inlet closer is usually the same as the outlet closer. However, such columns are entered and exited by a 1/16 inch o.d. tube and they cannot be used for syringe injection. With syringe injection it is important to ensure that the syringe needle does not penetrate the frit and disturb the packing. To prevent this the frit is topped either by a 5 mm layer of 200 μm glass beads into which the injection is made or by a porous Teflon frit inserted into the column. However, the authors have experienced problems with both systems when chromatographing very polar molecules due to adsorption on the glass beads or on to the Teflon. In this case it was found necessary to use a steel frit and to use a needle guide to ensure that the needle could not penetrate the frit.

Whichever end closers are used it is essential that the user has easy access to the frits, especially at the inlet end. Pump seal debris and insoluble components may collect on the frit or on the first few millimetres of the column packing and eventually block it. Furthermore, the packing may settle in use and leave a void under the frit which is only detected when the column efficiency falls disastrously. The simple expedient of removing the top few millimetres of packing and replacing with fresh material will increase the length of life of a column considerably.

3.10.3 Guard Columns

A short guard column (5.0 cm × 4.6 mm i.d.) used before the analytical column will also increase the life of the analytical column, by retaining non-eluted components and particulate matter. Since the guard column also separates components the composition of its packing should be the same as the analytical column. For maximum efficiency a microparticulate packing is used in the guard column. However, since it will be necessary periodically to replace the packing in the guard column the use of such packings requiring slurry packing is not always expedient. Pellicular packings in the 37–44 μm range can be packed 'dry' by a fill and tap technique and are therefore more convenient to use for guard columns. There will be a slight decrease in efficiency ($\sim$ 5%) from the use of the pellicular column, but this is easily outweighed by convenience in use, particularly when 'dirty' samples are analysed, e.g. body fluids.

The guard column must be connected to the analytical column by zero dead-volume fittings.

3.11 Column Thermostats

Control of column temperature is important in liquid chromatography. The effect of temperature on retention times and reproducibilities is quite significant, especially when using the reverse phase and ion-exchange modes. In some separations an increase in temperature of 6°C can bring about a 30% reduction in retention time. Chemically bonded stationary phases may be even more sensitive to temperature changes.

Forced air ovens. The majority of commercial instruments are available with a forced air oven which provides electronic control of temperature from ambient to about 150°C (higher temperatures being unnecessary) with temperature stability of ±0.2°C or better. A purge of inert gas is sometimes provided to prevent fire hazard should a leak occur in the flow system. A disadvantage of some forced air ovens is that they are not so adaptable to changes in column length.

Column jackets. Columns may be heated using glass or stainless-steel jackets with the circulating fluid heated in an external constant temperature bath. The temperature of the bath can be maintained to better than ±0.01°C, giving a column temperature control of better than ±0.1°C. Such heating jackets can be made cheaply in the laboratory, and different sizes can be made for different column lengths.

3.12 Liquid Chromatography Detectors

A major development in high efficiency liquid chromatography has been the development of sensitive, on-line detector systems. Unlike gas chromatography, in liquid chromatography the physical properties of the sample and mobile phase are often very similar. This has led to the development of two basic types of detector for use in liquid chromatography involving:

(a) the differential measurement of a property common to both the sample and the mobile phase;

(b) the measurement of a property that is specific to the sample, either with or without the removal of the mobile phase before detection.

Examples of the first type of detector, also known as bulk property detectors, are the differential refractometer, conductivity and dielectric constant detectors. Solute property detectors which do not require removal of the mobile phase before detection include the ultraviolet absorption, polarographic, and radioactivity detectors, whilst the moving wire flame ionization detector (FID) and the electron capture (EC) detector both require removal of the mobile phase before detection.

Before discussing examples of the various types of detector it is necessary to look at certain parameters, the values of which may determine the suitability of a detection system for a particular analysis.

Lower limit of detection. This is the minimum amount of substance that can be detected, i.e. produce a response discernible above the noise level, and is normally taken to be the concentration that produces a signal having twice the background noise level. Noise level refers to the peak-to-peak amplitude of random fluctuations about a base-line. Only noise that can be mistaken for a peak is significant in considering the limit of detection. Thus noise peaks, having the same frequency (width) as the sample peaks, which may appear as random peaks and valleys on the base-line, affect the limit of detection whereas high frequency noise which only broadens the base-line does not.

Some sources of noise are inherent in the detector; others such as temperature changes, the chemical composition of the mobile phase, the flow rate, and pressure are within the control of the analyst.

Ambient temperature changes may cause a base-line drift and are particularly noticeable when using refractometers.

Changes in the mobile phase composition will produce large base-line fluctuations, the most serious problems arising when the composition is deliberately changed as in gradient elution.

Flow rate and pressure changes may also cause random noise, especially in refractometers. Much of the noise due to these effects can be avoided in differential detectors by by-passing the reference cell and using a static reference rather than a flowing one.

The overall limit of detection is a function of both the detector and the entire chromatographic system, and depends on the amount of dilution (zone spreading) caused by the injector, column, connecting tubes, and detector. Typically, dilution factors vary from 5 to 100, depending largely on the diameter of the column, since narrow bore columns dilute the sample less than wide bore columns.

Sensitivity. This is the increase in detector signal per unit increase in solute concentration. To compare the sensitivity of two detectors it is necessary to define the noise level.

Linearity. The linear range of the system, where the detector signal is directly proportional to the concentration of the solute, is important in quantitative analysis. If sample concentrations are too high for the linear range, an appropriate dilution may be made or a calibration curve must be constructed.

Band broadening. Band broadening in the detector will add a variance to the peak width. To keep this to a minimum low volume detector cells and low volume connections are essential. This places certain limits on the cell geometry. For example, the absorbance of a u.v. detector is proportional to the optical path length and the sample concentration. The requirements of small volume and large optical path length necessitate a cell geometry in which the light passes coaxially through a long, small diameter cell with the direction of liquid flow. Equally important as the volume of the cell itself, can be the relative volume of the cell to the volume of the solute zone. It has been shown that this ratio should be less than 1 : 5, and preferably 1 : 10, for k' values of less than 2. For higher k' values the ratio is not critical.

The characteristics of an ideal detector for HPLC include the following. The detector should:

1. have a high sensitivy — a usable sensitivity of better than 0.1 μg of sample in 1 cm^3 of mobile phase;
2. respond universally to all solutes, or have a predictable specificity;
3. have a linear response over several orders of concentration
4. posses a low dead volume;
5. be non-destructive;
6. be insensitive to changes in temperature and mobile phase velocity changes;
7. operate continuously;
8. be reliable and convenient to use.

Unfortunately no single detector satisfies all these criteria, but as far as commercially available detectors are concerned three types find the most widespread use.

3.12.1 Photometric Detectors

Ultraviolet Detectors The most widely used detectors are those based on the absorption of ultraviolet light. They are therefore not universal in application, but a great many substances do absorb u.v. radiation including all substances having π-bonding electrons and also those with unshared (non-bonded) electrons, e.g. olefins, aromatics, and compounds containing $>C=O$, $>C=S$, $-N=O$, and $-N=N-$.

The radiation source used in many u.v. detectors is a low-pressure mercury vapour lamp. The predominant line in the spectrum is at 254 nm, and other lines are filtered out to give monochromatic light at this wavelength. Most u.v. absorbers at this wavelength have extinction coefficients in the range $10^2 - 10^{-1}$ dm^3 mol^{-1} cm^{-1} and this wavelength is suitable for a general purpose detector. Using a medium-pressure mercury source and the appropriate filters, operation at 254, 280, 313, 334 and 365 nm is possible. Variable wavelength u.v.–visible spectrophotometers use a deuterium lamp for the range 190–400 nm and a quartz–iodine or tungsten–halogen lamp in the range 350–700 nm. Variable wavelength detectors offer several advantages over a fixed wavelength instrument: (i) higher absorbance for many components and therefore increased sensitivity; (ii) greater selectivity, since a wavelength can be chosen where the solutes of interest absorb while others do not; and (iii) improved performance in gradient elution through the ability to select a wavelength where the solvents have the same absorption.

The cell volume in u.v. detectors is typically 8 μl and the cell 1 mm diameter $\times$ 10 mm long although smaller (4 μl) cells are available.

The ability to stop the mobile phase flow and to carry out a spectral scan on the solute held in the detector greatly enhances the analytical capability of the detector. To obtain the most meaningful results the spectral scan should be corrected for background solvent absorption and for variations in the lamp output. This is achieved by the addition of a memory module, either as a separate unit or as part of a general data system, which can store the mobile phase background spectrum and lamp characteristics and make the necessary corrections to the scan, thus giving improved peak resolution.

On one commercially available detector it is possible, simultaneously, to measure and to record the absorbance at two different wavelengths, or alternatively to monitor simultaneously at the same wavelength but at two different sensitivities. This latter facility obviously has advantages when a quantitative analysis of a sample consisting of a high proportion of one component and only small amounts of other components is required.

A problem in absorbance detection is that of distinguishing between a genuine absorbance due to the solute, and an absorbance due to refractive index changes in the sample cell [1]. If the refractive index of the cell changes, the amount of u.v. radiation reaching the detector will change, because more radiation will strike the cell walls and will be adsorbed. This has been overcome in the Waters Associates

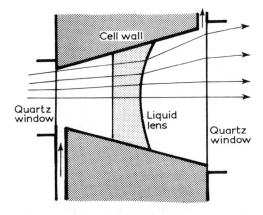

Fig. 3.15 Taper cell (after Waters Associates).

taper cell (Fig. 3.15) in which u.v. light striking the quartz window of the cell is refracted but does not strike the walls of the cell because of the taper. Similarly, if the light strikes a region of differing refractive index, which acts as a liquid lens within the cell, the taper again reduces the amount of adsorption at the cell wall. Many detector cells can only be operated at pressures up to 500 lbf in^{-2} and if it is required to use the detector in a differential mode a dual column system must be employed. However, cells are available that will withstand pressures up to 2000 lbf in^{-2} and allow the detector reference cell to be connected directly to the solvent pump, the reference cell output then going via the injecting system and column to the sample cell without the use of a pressure dropping reference column.

Fig. 3.16 shows a schematic diagram of a dual-beam u.v. absorbance detector with the taper cells. Light from the u.v. source is passed through a defining aperture or a phosphor converter to give the required wavelengths (λ_1 and λ_2). The lens focuses the light so that it is parallel when it strikes the cell. Having passed through the reference and sample cells the light passes through a window and filter before striking the detector photocell. The signals generated are

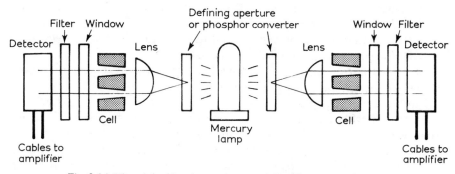

Fig. 3.16 Ultraviolet absorbance detector (after Waters Associates).

proportional to the intensity of the light. These signals are pre-amplified and then fed to a log ratio circuit to generate a log function. The display can be recorded as two separate traces at two different wavelengths (λ_1 and λ_2), as two traces at the same wavelength but different sensitivities, or in a differential mode ($\lambda_1 - \lambda_2$) giving a single trace.

Ultraviolet absorption detectors are relatively insensitive to temperature and flow changes and have a high sensitivity to many samples. With sensitivities of 0.001 absorbance unit, full scale deflection, and noise levels typically 1%, it is possible to detect as little as 1 ng of a solute with a moderate u.v. absorbance. The wide linear dynamic range ($\sim 10^4$) of these detectors makes it possible to measure both trace and major components on the same chromatogram.

Since these detectors are measuring a solute property rather than a bulk property they are ideal for use with the gradient elution technique. However, changes in the u.v. absorption as the solvent composition changes may cause a drifting baseline which can give rise to distorted peaks. This too may be overcome by the use of a data system or memory module as for a spectral scan.

Visible Photometers. Few compounds absorb in the visible region of the spectrum. However, reagents may be added to the effluent either continuously or in a regular manner to form suitable coloured compounds which can be monitored by absorption photometry. This technique is used in the automatic amino acid analysers where the purple colour, obtained by reaction of an amino acid with ninhydrin, is monitored at 570 nm. Some colour-forming reagents are fairly general in their reaction whilst others are very specific, so the technique may be used either as a general or a specific detector. In either case the colour-forming reaction must be fast, and the volumes of reactants involved small, so that the reaction does not contribute to the peak width.

Infrared Photometers. Absorption in the infrared region can be used both as a general and as a specific detector. For use as a general detector a stopped-flow technique must be used. In this mode the sample can be held in the detector cell and the wavelength scanned to detect various functional groups. Alternatively, the sample may be trapped out and analysed. A suitable trapping system has been described by Juhasz *et al.* [2].

As a specific detector the wavelength is set at a particular value for the detection of a single functional group. Some examples are given in Table 3.2. Of course, a detector operating at 3.4 μm will respond to any organic compound exhibiting a C−H absorption and is essentially a universal detector.

One such instrument is the Miran Detector (Foxboro/Wilks, Inc.). This is a single-beam spectrometer with a continuously variable filter which scans from 2.5 to 14.5 μm (4000-690 cm^{-1}), with a resolution of 0.12 μm at 5 μm. Thus both stopped-flow with scan and single-wave-length monitoring is possible. The low volume flow cells (5-200 μl, pathlength 1.5 mm) can be thermostatted up to 200°C for exclusion chromatography. However, the detector is not sensitive to temperature or flow-rate variations. The detector can be used with gradient elution as shown by the separation of a synthetic wax mixture (Fig. 3.17) although base-

Table 3.2 *Infrared adsorption frequencies*

Compound type	Functional group	Wavelength/μm
Aliphatic hydrocarbon	$-C-H$	3.40
Aromatic hydrocarbon	⌬	3.20
Olefin	$>C=C<$	6.10
Alcohol	$-CH_2OH$	3.00
Carbonyl	$CH_3-C{<}^{O}_{O-R}$	5.75
Ketone	$R-C-R$ ‖ O	5.80

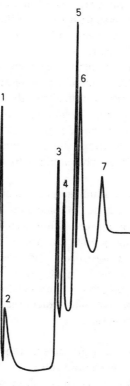

Fig. 3.17 Analysis of a synthetic wax mixture—detection with an infrared detector at 5.75 μm.

1. Octadecyl decanoate   2. Methyl stearate   3. Tridecan-2-one
methyl undecyl ketane   4. Tripalmitin   5. Octaldehyde
        6. 1, 2-Dipalmitoyl glycerol   7. Glycerol monostearate

line shifts are experienced.

One of the major problems with the infrared detector is the choice of mobile phase particularly when using the scan mode. Since i.r. detectors measure absorbance at selected wavelength the availability of 'spectral windows' in the spectrum of the mobile phase must be considered as well as the usual chromatographic requirements. A spectral window is commonly defined as a region of the spectrum where the solvent, in a sample cell of given path-length, shows at least 30% transmittance. Thus solvents in a longer cell will have fewer windows e.g. with a 1 mm path-length cell acetonitrile has a window from 6–8 μm and from 8.5–11 μm, however both of these regions are spectrally opaque in a 3 mm cell. This suggests that short path-length cells are advantageous. Indeed at a path-length of 0.1 mm. most common chromatographic solvents have substantial windows but the shorter path-length reduces the sensitivity of the detection.

Since the magnitude of the absorption varies from one solvent to another and from one wavelength to another base-line shifts are usually observed during gradient elution. These base-line shifts are often linear with concentration of the second solvent. The size of such shifts depends on the extent of the compositional change and on the absorbance characteristics of the mobile phase, the optical path-length and the detector sensitivity.

The principle of absorbance matching may be applied to i.r. detection to reduce these base-line shifts. Absorbance matching involves the blending of solvents to give mobile phase pairs with similar optical absorbance for use in gradient elution. For example, using acetonitrile as the primary solvent and a secondary solvent consisting of a mixture of methylene chloride and tetrahydrofuran that closely matches the absorbance of acetonitrile, carbonyl-containing compounds may be eluted under gradient conditions with minimal base-line shift, and can be monitored at 5.8 μm.

Fluorescence Photometers. Fluorescence spectroscopy can be used as a detection method with a high degree of selectivity and sensitivity for fluorescing compounds. Compounds which show fluorescence naturally are those with a conjugated cyclic structure, e.g. polynuclear aromatics, aflatoxins, aromatic amino acids, phenols, quinolines and oestrogens. However, the technique may be extended to a wider range of chemical types by synthesizing fluorescent derivatives. Two common derivatizing agents for this purpose are dansyl chloride (5-dimethyl-aminonaphthalene-1-sulphonyl chloride) which reacts with primary and secondary amines and phenols, and fluorescamine which reacts with primary amines and biogenic amino acids.

If a molecule or atom absorbs light the electrons are excited and a number of processes may occur before the electrons return to the ground state, e.g. internal conversion, quenching, phosphorescence and fluorescence. If the electrons return to the ground state is almost instantaneous and energy in the form of light is emitted, it is said to fluoresce. Because there will be some loss in energy the emitted light will be at a longer wavelength than the absorbed light.

The sensitivity of fluorescence measurements arises from the fact that the background emission is virtually zero so that large changes in emission are observed. This contrasts with the normal absorption where the difference between the incident and transmitted radiation can be quite small.

The high degree of selectivity arises from the fact that two wavelengths must be selected, that of the excitation energy and the emission energy. Careful choice of the two wavelengths can give extra selectivity.

The excitation radiation, 190–400 nm (deuterium lamp) and 350–600 nm (tungsten lamp), is either filtered or monochromated to give the desired excitation radiation, is focused onto a low-volume flow cell (5–30 μl) and the fluorescent emission is filtered to remove unwanted radiation and is detected by a photomultiplier. The fluorescent emission is usually measured at 90° to the excitation beam. However, since the emission is in all directions, this inevitably leads to a loss in sensitivity. The problem has been overcome in the Kratos FS970 detector by using a hemispherical collector. This, and a general schematic are shown in Fig. 3.18.

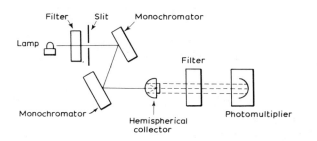

Fig. 3.18 Schematic diagram of a fluorescence detector – Kratos F.S. 970

The choice between an instrument which uses filters to select the incident wavelength or a monochromator depends on three factors; sensitivity, selectivity and cost. Filter fluorimeters give greater sensitivity as more light can be transmitted and they also cost much less than monochromator instruments. Monochromator instruments, however, give greater selectivity as the bandwidths of excitation and emission are narrower.

The fluorescence detector is some 100 times more sensitive than the absorption detector and picogram quantities of a favourable sample, e.g. anthracene, may be detected. The linear dynamic range is $\sim 10^3 - 10^4$ but it becomes non-linear at high concentrations due to fluorescence quenching as a result of energy absorption.

3.12.2 *Refractometers*
The differential refactometer is the second most widely used LC detector. It is in common use for preparative separations and in exclusion chromatography as well as in analytical HPLC. This detector continuously monitors the difference

in refractive index between the pure mobile phase and the mobile plus sample as it elutes from the column.

The main advantage of refractometers is that, since they are bulk property detectors, they are of universal application; the only requirement is that there is a difference in refractive index between the sample and the mobile phase, and this can be as little as 10^{-7} refractive index units (r.i.u.). The sensitivity of the detector depends upon this refractive index difference, but even under optimum conditions it is about two orders of magnitude less sensitive than the u.v. photometer. The sensitivity, defined as being equal to the noise level, is generally of the order of 10^{-7} - 10^{-8} r.i.u. It is a good approximation to assume 10^{-7} r.i.u. corresponds to 1 p.p.m. of sample in the eluent [3]. Refractometers are very sensitive to uncompensated changes in ambient temperature and rely mainly on the reference cell for this compensation. It is estimated that the reference cell compensates for 98–99% of an ambient temperature change [4], and the detectors therefore require an environment stable to between 0.001 and 0.01°C. This is achieved by circulating water from a temperature regulated supply through a metal block which contains the refractometer. Even so, to achieve optimum sensitivity and stability, efficient heat exchangers between the column outlet and the detector cell are required. If properly designed these heat exchangers do not add significantly to the peak broadening, and noise levels of 3×10^{-8} r.i.u. can be achieved [3].

If ambient temperature fluctuations $<\pm 0.5°C$ the detector may be run without water circulation but with static water in the jacket. The inlet and outlet of the water jacket should either be left open or be plugged—not connected together. If connected together water may circulate by convection and cause temperature fluctuations.

The reference cell is usually used in the static mode. Again it is important not to connect the inlet and outlet tubes together or again convectional flow may occur and cause temperature fluctuations.

The flow cell of the refractometer normally reaches an equilibrium temperature a few degrees above ambient [5]. At high flow rates the incoming mobile phase has less time to equilibrate with the flow cell temperature. This results in a flow rate dependent mobile phase temperature which, because of the temperature coefficient of refractive index, results in flow sensitivity. Pulsating flow manifests itself as base-line noise whilst slow variations in flow rate cause base-line drift. However, if the detector is properly thermostatted it becomes relatively insensitive to carrier flow, and flow programming even becomes possible.

The refractometer cells are less pressure resistant than u.v.-detector cells and back-pressure regulator should never be used. To avoid the formation of bubbles in the detector mobile phases should be pre-mixed *then* degassed. If used in series with another detector the refractometer should be placed last to avoid back pressure.

Because the detector is sensitive to changes in mobile phase composition it is not really practical to use it with gradient elution. By matching the refractive indices of the two end-points of eluting gradients, and by using a dual pump and column chromatograph, gradient elution with the refractometer has been shown to be feasible [6] but this is not generally applicable.

There are three types of refractometer cell in use. The first uses the principle of measurement based on Fresnel's law, which relates the transmittance and reflectance at a dielectric interface to the angle of light incidence and the refractive indices of the two materials forming the interface. In the refractometer the interface is formed between a glass prism and the liquid whose refractive index is to be measured.

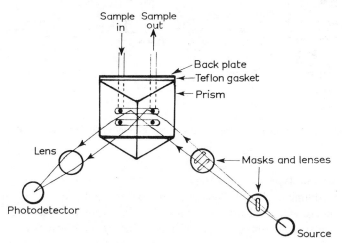

Fig. 3.19 Fresnel type refractometer (after Perkin-Elmer).

Fig. 3.19 shows one such Fresnel detector. Light from the source lamp is collimated into two beams by masks and a lens, enters the cell prism, and is focused on the glass—liquid interfaces formed by the sample and reference liquids which are in contact with the prism. Coarse and fine controls are provided for the adjustment of the angle of incidence on the interfaces. The sample and reference chambers are in a thin Teflon gasket (0.001 inch thick) clamped between the prism and a stainless steel back plate which contains the inlet and outlet tubes. The light, having passed through the two interfaces in the cell, passes through the thin liquid film and is reflected by the back plate. After passing out of the prism it is focused by lenses on to a dual element photodetector and the electrical signals produced are compared electronically and may be displayed on a normal millivolt recorder.

A disadvantage of the Fresnel detector is that to cover the normal refractive index range (1.31—1.55) two prisms are required. The flow cells commonly have a volume of 5 μl or less and are often favoured for use with very high performance LC columns. The small volume also minimizes dilution and therefore maximizes sensitivity, with the result that sensitivities of 1 μg cm^{-3} (sucrose in water) are obtainable. The flow-through characteristics of the cells help to reduce contamination problems but for proper operation the cell windows must be kept clean.

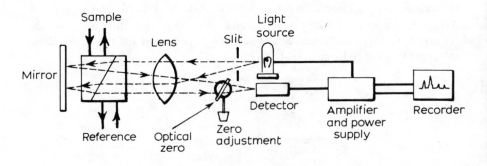

Fig. 3.20 Schematic diagram of a refractometer (after Waters Associates).

The second type of refractometer is shown in Fig. 3.20. Light from the source is collimated by the lens and falls on the cell. The cell consists of the sample and reference chambers separated by a diagonal sheet of glass. As the light passes through the cell it is refracted, then reflected from the mirror and refracted again as it passes back through the cell. The light is again focused by the lens, and the final position of the light beam as it strikes the detector will depend on the difference between the refractive index of the sample and that of the reference. The detector gives an electrical signal, proportional to the position of the light, which is amplified and can be displayed on a millivolt recorder.

This type of detector needs only one cell to cover the whole refractive index range normally required, and the cells are less susceptible to contamination than are those of the Fresnel type.

The third type of refractometer uses the shearing interferometer principle for measurement. This is shown schematically in Fig. 3.21. The beam from the light source (546 nm) is split into two parts by the beam splitter, is focused by the lens, and passes through the sample and reference cells (15 μl volume with a 10 mm path length; other cells are available for preparative use). The light beams are recombined by a second lens and beam splitter and fall on the interferometer detector. A difference in refractive index between the sample and the reference produces a difference in optical path length. This is measured by the interferometer in fractions of the wavelength of light; therefore no calibration is required and the response is completely linear. The sensitivity, defined as equal to the noise

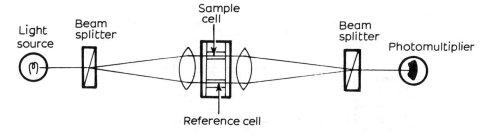

Fig. 3.21 Shearing interferometer type refractometer (after Optilab).

level, is 5×10^{-9} r.i.u. corresponding to 0.1 μg, or less, of sucrose. This type of refractometer undoubtedly gives the highest sensitivity and its sensitivity is independent of solvent refractive index (n)—for the other types it is proportional to $1/n$. Its calibration is also absolute. However, it is considerably more expensive.

It is probably true to say that the refractometer has reached its limit as regards sensitivity, but stability is now better allowing easier use of the maximum sensitivity range.

3.12.3 Solute Transport (Phase Transformation) Detectors

Although these detectors are not commercially available at the moment they are in use and there is still some, though as yet unsatisfied, demand for them. In particular the moving-wire detector is in demand for the analysis of non-u.v.-absorbing lipids.

Moving-Wire Detectors. These detectors are based on the flame ionization detector (FID) and are therefore of general application. The column effluent is fed continuously on to a transport system, such as a moving wire, disc, chain, or helix. The solvent is evaporated in a furnace and the non-volatile sample passes into an FID for detection. It appears therefore that this type of detection system should possess most of the attributes of an 'ideal' detector: high sensitivity, universal response, independence of the chromatographic process, and a wide, linear dynamic range. In practice, however, they possess several disadvantages which have hindered their acceptance as LC detectors at present; these include a high cost, bulkiness, relatively poor sensitivity, and operational inconvenience. Poor sensitivity is due to the relatively small amount of solute that reaches the detector on the transport system. Furthermore, extra-column band broadening can be significant. Quantitative results with the detector are poor because the amount of sample 'sticking' to the transport system will depend on the surface tension of the solute and solvent and on the wire speed.

A schematic diagram of a wire transport detector is shown in Fig. 3.22. Stainless-steel wire from the feed spool is passed through a glass tube which is heated to 850°C and purged with a flow of air. This cleans the wire and oxidizes organic contaminants. The wire passes round a large horizontal pulley and is

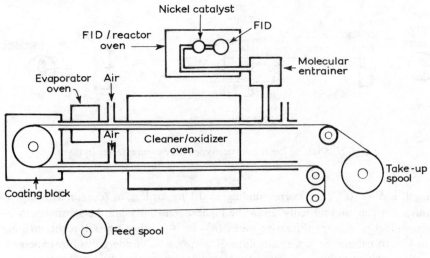

Fig. 3.22 Schematic diagram of a moving wire FID transport detector
(after Pye Unicam Ltd.).

coated with column effluent. It then passes through an evaporator oven, the temperature of which is controllable between ambient and 350°C, and the solvent is removed. The solute remaining passes through a second oven (at 700–800°C) with an oxygen purge where the solute is burned to carbon dioxide and water. The carbon dioxide is drawn into a molecular entrainer, hydrogen is added, and the whole mixture is passed over a nickel catalyst held at 330°C. This reduces the carbon dioxide to methane which is then passed by the gas stream to the flame ionization detector, the response of which will be proportional to the number of carbon atoms in the solute molecule.

The attributes of the FID are well known, but used in this manner its full potential cannot be realized because of deficiencies in the transport system. However, several authors [7–10] have reported improved sensitivity by modifications of the transport system, and the considerable potential of these detectors may yet be realized.

Electron Capture Detector. The electron capture detector (ECD) is widely used in gas chromatography, together with the nitrogen–phosphorous detector, for the analysis of pesticides in the environment. Both of these detectors combine a high degree of selectivity with a high level of sensitivity.

Electron capture detection has been combined with HPLC by vaporizing the total eluent from the LC column and passing it directly into an ECD (^{63}Ni electron capture detector by Pye Unicam Ltd.) [11]. The eluent is vaporized by passing it into a stainless-steel transfer tube mounted in the detector oven and maintained at 300°C. A nitrogen purge sweeps the vapour through the detector into a stainless-steel condenser coil from which it can be collected as a liquid.

The sensitivity of this detector to organochlorine pesticides separated by HPLC is lower than that achieved when the separation is by GC, but this is partly because of an increase in background noise due to electron-capturing species in the solvent vapour. However, the system does show a significant increase in sensitivity over all other LC detectors for this type of analysis.

Choice of solvent will be limited when using the ECD since common solvents such as chloroform and carbon tetrachloride cannot be used. Many polar solvents too must be diluted with saturated hydrocarbons if the standing current is not to be reduced to an unacceptable level. The presence of dissolved oxygen in the solvent causes an increase in noise level and a reduction in the standing current. Oxygen can be removed by purging the solvent with oxygen-free nitrogen prior to use. The solvents should also be free of electron-capturing impurities.

The ECD is specific for electrophilic species such as compounds containing halogens, phosphorus, sulphur, lead, oxygen, etc., and finds application wherever these species are to be found at very low concentration levels (<1 p.p.b.).

3.12.4   Radioactivity Detectors

Detectors used for measuring the radioactivity of solutes separated chromatographically are based on standard Geiger counting and scintillation systems. Sensitivities obtainable with flow-through systems are lower than those obtained when counting individual samples from a fraction collector. The sensitivity can be increased by increasing the counting time, but this can only be achieved by a lowering of mobile phase flow rates, or by using larger volume flow cells so that the residence time of the solute in the detector is increased. However, both of these possibilities lead to lower chromatographic efficiencies. A stopped flow technique, however, allows for larger counting times without destroying the chromatographic separation. Radiochemical methods of detection are therefore best suited for use with relatively large columns and long analysis times.

3.12.5   Electrochemical Detectors

Electrochemical methods of detection possibly offer the most promising approach to the problem of developing a universal detector for HPLC. They offer the benefits of high sensitivity and selectivity and are well suited to trace analysis.

Electrochemical detectors have developed along several lines; polarography, amperometry, coulometry and conductivity, but their commercial availability is still rather limited.

Early attention was given mainly to the use of polarographic detectors with a dropping mercury electrode (DME) [12]. However, although the DME gives a reproducible electrode surface and large cathodic range, its limited anodic range restricts its use to electro-oxidizable systems. The use of a horizontal mercury capillary [13] has been applied to an improved DME detector [14].

Solid electrode sensors [15, 16], a wall-jet electrochemical detector [17] and one which allows the use of both solid sensors and a DME have been described [18].

The use of amperometric and coulometric detectors for HPLC has been reviewed by Kissinger [19].

An *amperometric detector* (Chromatix) is available commercially. This is suitable for detecting electroactive compounds from ion-exchange and reverse phase separation modes with aqueous mobile phases. Eluting compounds are oxidized or reduced at a carbon/polymer electrode under a constant applied potential. The resulting current flow at the electrode, proportional to the sample concentration, is converted into voltage output and is monitored as a function of time. By changing the potential range selectivity can be introduced, and non-electroactive species are not detected.

An electrolytic conductivity detector has been available for gas chromatography for some time [20]. A *photo-conductivity* detector based on the development by Rodgers and Hall [21] is now available commercially (Tracor 965 Photo-conductivity Detector, Tracor Inc.) [22].

The *photo-conductivity* detector uses post-column photochemical reactions to produce ionic species which, when dissolved in an electrolyte, are detected in the conductivity cell. The effluent from the column is split into two streams; analytical and reference. On the analytical side the sample passes through a quartz reaction coil where it is irradiated by u.v. light and then passes into the conductivity cell. The balance of the effluent passes first into a reference delay coil whose flow characteristics match the reaction coil and then to the reference side of the conductivity cell. The difference in conducitivity measures the photolysis products. Thus both ionic and photo-ionizable species are detected and the detector is applicable to the determination of a wide range of compounds containing certain halogen, nitrogen and sulphur groups at the picogram level. Furthermore, it is relatively insensitive to aromatics and other u.v.-absorbing compounds lacking a heteroatom.

3.12.6 Mass Spectrometers

Liquid chromatography–mass spectrometry (LC–MS) is now a commercial reality. We shall only concern ourselves with the LC–MS interface since a consideration of the mass spectrometer itself is beyond the scope of this book.

The earlier system took a proportion of the LC effluent directly into the ion-source *via* a capillary inlet [23]. The volatilized solvent then became a chemical ionization (CI) reactant gas and the sample was ionized by proton transfer from the ionized solvent molecules. This system only allowed the use of low mobile phase flow rates (10–20 μl min^{-1} compared with the more usual 1.0 cm^3 min^{-1}), only solvent CI spectra could be obtained and cryogenic source pumping was desirable.

The method now adopted involves an endless moving belt interface [24]. The effluent from the liquid chromatograph is introduced directly onto a continuously moving belt, made from stainless steel or Kapton (polyimide). The solvent is partially evaporated by passing under an infrared heater and solvent evaporation is completed when the sample passes through two pumped vacuum locks and into the spectrometer source. Inside the spectrometer vacuum manifold the belt enters a flash vapourizer attached to the ion source. As the belt leaves the ion chamber it is again heated to remove any residual sample that might give a memory effect as the belt recycles.

In this way the interface is able to maintain a high sample yield and yet remove the solvent and also to volatilize the sample without decomposition.

Actual LC flow rates that the interface can accept depend on the volatility and polarity of the solvent. Volatile, non-polar solvents can be accepted up to 2 cm min^{-1}, whereas water must be below 0.3 cm min^{-1}. The efficiency of the transfer is also dependent on the nature of the mobile phase and of the sample, but these are typically between 30% and 80%.

The LC–MS interface does not affect the usual ionization methods and both chemical ionization (CI) and electron ionization (EI) may be used. As in GC–MS the LC–MS combination can be used either to scan the full mass range of interest or to perform selective ion monitoring. The LC interface can also be used for probe samples where its ability to handle thermally labile and involatile samples is an advantage over conventional inlets.

3.12.7 Summary of Detector Characteristics

Because of its high sensitivity, reproducibility, and ease of operation, the u.v. detector will be the natural choice for many analyses. For non-u.v. adsorbing solutes the refractive index detector is the most probable choice.

Table 3.3 summarizes the characteristics of the most widely used detectors for HPLC, and consideration of these will assist in the selection of the most suitable detector for a particular application.

3.13 Flow Rate Measurement

Flow rate measurement is necessary to provide precise retention data. It is also useful as an advanced warning system since mobile phase flow variations can indicate a failure in the pumping system or a decrease in column permeability due to packing failure or a partial blockage.

The feed-back flow controllers referred to in Section 3.3 can be adapted to provide a continuous read-out of the flow rate, making it possible to compare directly chromatograms run at different flow rates.

More conventional, and much cheaper, methods are (a) the collection, in a calibrated vessel, of the mobile phase for a known length of time and measurement of the volume collected; (b) the collection for a known length of time followed by weighing the amount of mobile phase collected; and (c) the flow-tube method, which involves introducing an air bubble into the mobile phase stream as it passes through a calibrated, transparent, horizontal tube connected to the detector outlet, and measuring the time for the air bubble to travel between two marks representing a known volume (say 1.0 or 2.0 cm^3).

Methods (a) and (c) are most commonly used; they give a precision better

Table 3.3 *Typical characteristics of detectors* [after L. R. Snyder and J. J. Kirkland, Introduction to Modern Liquid Chromatography, (2nd Edition) Wiley, New York, 1974].

	UV (absorbance)	RI (r.i. units)	Transport		Radioactivity	Polarography (µamp)	IR (absorbance)	Fluorimeter	Conductivity (µmho)
			FID (amp)	EC (amp)					
Response	specific	general	general	specific	specific	specific	specific	specific	specific
Use with gradient elution	yes	no	yes	yes	yes	N.A.	yes	yes	no
Linear dynamic range (upper limit)	2–3	10^{-8}	10^{-8}	N.A.	N.A.	2×10^{-5}	1.0	N.A.	1000
Linear range	10^5	10^4	$\sim 10^5$	5×10^2	large	10^6	10^4	$\sim 10^6$	2×10^4
Sensitivity at 1% noise, full-scale	0.002	2×10^{-6}	10^{-11}	N.A.	N.A.	2×10^{-6}	0.01	0.005	0.05
Sensitivity to favourable sample	2×10^{-10} g cm^{-3}	5×10^{-7} g cm^{-3}	$\sim 5 \times 10^{-7}$ g cm^{-3}	10^{-10} g cm^{-3}	50 c.p.m ^{14}C cm^{-3}	10^{-12} g cm^{-3}	10^{-6} g cm^{-3}	$\sim 10^{-12}$ g cm^{-3}	10^{-8} g cm^{-3}
Flow sensitive	no	no	yes	yes	no	yes	no	no	no
Temp. sensitivity	low	10^{-4}°C	negligible	negligible	negligible	1.5%/°C	low	low	2%/°C

N.A. – Figures not available.

than 1%. The gravimetric method (b) is capable of the highest precision but is tedious and is usually used only to calibrate pump flows.

3.14 Fraction Collectors
Because of the relatively short analysis times in modern liquid chromatography, manual collection of samples can conveniently be made. A manually operated low dead volume valve on the output side of the detector allows for sample collection, diverting the effluent to waste, or returning the mobile phase to the solvent reservoir for re-use.

In analytical gel permeation chromatography and in preparative scale operation, completely automatic fraction collectors are often used. Fraction collection is controlled by signals from a control module which is programmed to select individual peaks or fractions of peaks, according to the level of the peak signal on the chromatogram.

The control module can also be used to control the automatic injection of samples, thus making for a fully automatic preparative mode of operation.

3.15 Data Handling
Data handling in chromatography now ranges from a simple pen recorder to complicated computing integrators and computerized data handling systems.

Pen recorders should have a response time of less than 1 second for full-scale deflection and a wide range of chart speeds. Chart speeds generally used are considerably faster then in gas chromatography; a range from 0.5 to 30 cm s^{-1} is desirable.

In order to obtain precise quantitative data and obtain maximum benefit from computerized data handling, several factors must be controlled. The injector, column, detector, and data system must operate reproducibly, and the solvent flow rate and column temperature must remain stable during the run.

Developments in electronic calculator circuit-chip design have led to the development of micro-processor based computing integrators which, as well as providing peak detector and peak area measurement, are capable of controlling external equipment so as to provide control of these factors.

For successful data reduction in chromatography the integrator should have most of the following capabilities:

1. measurement of retention times;
2. automatic peak detection, peak area measurement, and base-line correction;
3. peak area allocation for unresolved peaks;
4. calculation of sample concentrations from peak area measurements;
5. component identification.
6. the ability to store raw data in memory so that programs can be edited and results recalculated without running the sample again;
7. convenient operator interface and interface to hierarchical computers for more complex calculations.

The first three of these capabilities are to be found on the simpler and cheaper digital integrators using analogue logic circuits, but the full capabilities are to be found only on the more expensive computing integrators. For the laboratory carrying out a large number of routine analyses the additional cost of the computing integrator is well justified, but for the chromatographer involved in 'one off' analyses the digital integrator may well give him all the information he requires.

3.16 Microprocessor Controlled HPLC

Several manufacturers today offer micro-processor controlled chromatographs. These fully automate analyses for unattended instrument operation. Thus the solvent delivery system, injector, oven, detector, fraction collector and data reduction can all be carried out under the control of a central micro-processor. With the capability to program sequential instrument parameters such as flow rate, mobile phase composition and detector wavelength the development of a separation can be carried out unattended.

Certainly the use of micro-processors has increased the precision of flow rate control, gradient elution, injection and several other aspects of the chromatographic system but the cost is high. Instrument time can be more than doubled by unattended use, though the authors know the frustration of finding a pile of worthless paper in the morning! Furthermore, in spite of the ingenuity of the manufacturers it is still only possible to do one separation at a time and several more simple, cheaper systems may offer more flexibility and ultimately a greater analytical capability.

References
1. Little, J. N. and Fallick, G. J. (1975) *J. Chromatog.*, **112**, 389.
2. Juhasz, A. A., Doali, J. O. and Rocchio, J. J. (1974) *International Laboratory*, July/August, 21.
3. Deininger, G. and Halasz, I. (1970) *J. Chromatog. Sci.*, **8**, 499.
4. Munk, M. N. (1970) *J. Chromatog. Sci.*, **8**, 491.
5. Watson, E. S. (1969) *American Laboratory*, September.
6. Bombaugh, K. J., King, R. N. and Cohen, A. J. (1969) *J. Chromatog.*, **43**, 332.
7. Karmen, A. (1966) *Anal. Chem.*, **38**, 286.
8. Labidus, B. M. and Karmen, A. (1972) *J. Chromatog. Sci.*, **10**, 103.
9. Stolyhwo, A., Prevett, O. S. and Erdahl, W. L. (1973) *J. Chromatog. Sci.*, **11**, 263.
10. Pretorius, V. and Rensburg, J. F. J. (1973) *J. Chromatog. Sci.*, **11**, 355.
11. Willmott, F. W. and Dolphin, R. J. (1976) *Spectrophotometry Chromatography Analytical News*, 7, 6.
12. Fleet, N. and Little, C. J. (1974) *J. Chromatog. Sci.*, **12**, 747.
13. Scarano, E., Bonicelli, M. G. and Forina, M. (1970) *Anal. Chem.*, **42**, 1470.
14. Hanekamp, H. B., Bos, P. and Frei, R. W. (1979) *Advances in Chromatography*, (ed. A. Zlatkis), Houston, U.S.A.
15. Adams, R. N. (1969) *Electrochemistry at Solid Electrodes*, Dekker, New York.

16. Alder, J. F., Fleet, B. and Kane, P. O. (1971) *J. Electroanal. Chem.*, **30**, 427.
17. Yamada, J. and Matsuda, H. J. (1973) *Electroanal. Chem.*, **44**, 189.
18. Beachamp, R., Boinay, P., Fombon, J.-J., Tacussel, J., Breant, M., Georges, J., Porthault, M. and Vittori, O. (1981) *J. Chromatog.*, **204**, 123.
19. Kissinger, P. I. (1977) *Anal. Chem.*, **49**, 477.
20. Hall, 700A Electrolytic Conductivity Detector Tracor Inc. (1978).
21. Rodgers, D. H. and Hall, R. C. (1977) Pittsburgh Conference of Analytical Chemistry, Paper No. 8.
22. Popovich, D. J., Dixon, J. B. and Ehrlich, B. J. (1979) *J. Chromatog. Sci.*, **17**, 643.
23. Arpino, P. J., Dawkins, B. G. and McLafferty, F. W. (1974) *J. Chromatog. Sci.*, **12**, 574.
24. McFadden, W., Schwartz, H. L. and Evans, S. (1975) *J. Chromatog.*, **122**, 389.

4

Stationary phases in liquid chromatography

4.1 Introduction

Although HPLC has only been available for fifteen years, the column part of the system has seen more changes than any other part of the system. Classical column chromatography was associated with large diameter, porous particle packings ($>100 \mu m$) in wide bore columns (1–2 cm i.d.) and a low head pressure (from gravity to 50 lbf in^{-2}) which together lead to slow analyses. The large diameter porous particles are unsuitable for HPLC owing to slow diffusion into the deeper pores which causes a decrease in column efficiency and loss of resolution. The earlier column packing materials also had very irregular shapes which formed in-homogeneously packed beds producing great variation in the mobile phase velocity and band spreading so that column efficiencies and resolution were again low. Equation (2.42) shows the relationship between efficiency in terms of height equivalents to a theoretical plate and many other variables:

$$H = C_D' \frac{D_M}{v} + C_S' \frac{d_f^2 v}{D_S} + C_{SM}' \frac{d_p^2 v}{D_M} + \frac{1}{1/C_F' d_p + D_M/C_M' v d_p^2} \qquad (2.42)$$

With LC regular adsorbents, d_f the film thickness of the stationary phase is equivalent to d_p since the porosity of the adsorbent allows the solute to diffuse through the whole particle. Since C_S, the resistance to mass transfer, is proportional to d_f^2, and since C_S contributes linearly to the plate height increase in the flow velocity for HPLC the particle diameter must be as small as possible.

Manufacturers then provided small diameter regular particles which meant that the solute molecules did not have to travel very far between the particles, giving rapid mass transfer. The resulting rapid equilibrium provides increased column efficiency. It is estimated that smaller porous particles give column efficiencies two orders of magnitude greater than the older large porous particles [1].

Today HPLC is associated with small diameter particle porous packings (down to 3 μm), narrow bore columns (down to 1 mm i.d.). and high inlet pressures (5000 lbf in^{-2}).

4.2 Stationary Phase Types

Two ways of producing small regular packing materials have been based on either microporous particles or porous layer beads (pellicular beads).

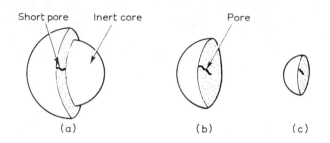

Fig. 4.1 Types of HPLC packings [8]. (a) Pellicular beads (30–45 μm) with short pores. (b) Microporous particles (20–40 μm) with longer pores. (c) Microporous particles (3–10 μm) with short pores.

4.2.1 Pellicular Beads

A major problem confronting the beginner in HPLC is the vast selection of phases available in the trade literature. Many of these packings have been listed in the tables to permit the beginner to see that many differently named products will do the same job.

The pellicular bead materials (also called superficially porous particles, controlled surface porosity, solid core or porous layer supports) first reported by Horvath *et al.* [2] consist of a solid, non-porous core (usually silica) with a thin porous outer shell (1/30 to 1/40 the diameter of the solid core). The outer shell can be silica, alumina, or an ion exchange resin. From Table 4.1 it can be seen that the overall diameter of these particles is approximately 40 μm. These particles have a distinct advantage in that they can be packed dry into the column; their major disadvantage is that their short pores have small surface areas $(1-25 \text{ m}^2 \text{ g}^{-1})$ with the result that the particles have little capacity. Whilst they provide a perfectly reasonable packing material when the quantity of solutes is small, they overload easily and are not satisfactory for preparative LC. The pellicular beads are rarely used today except in guard columns, but we have kept the list of phases in Table 4.1 so that the earlier literature references to these materials may be understood.

4.2.2 Microporous Particles

The microporous particles (also called small porous particles or microparticles)

Table 4.1 *Pellicular bead packing materials for liquid–solid and liquid–liquid chromatography* [1].

Type	Name	Suppliers	Particle size (μm)	Surface area ($m^2 g^{-1}$)	Loading for LLC opt.	Description
Silica (active)	Actichrom	4, 27	40	25	4	Glass powder; uniform surface activity; higher than PLB
	Corasil I & II	31	37–50	I 7, II 14	1	Corasil II has a double coating of silica; LLC loadings are twice those of Corasil I
	Pellosil HS, Pellosil HC	1, 4, 15, 20, 27, 29	37–44	HS 4, HC 8	1	HC has a thicker coating than HS (HC means high capacity, HS high speed)
	Perisorb A	11, 13, 25, 27, 29, 30	30–40	14	1	Pore volume 0.05 cm^3 g^{-1}
	Vydac	2, 5, 13, 16, 22, 26, 27, 29, 30	30–44	12	2	Average pore diameter 57 Å; avoid solvents more polar than methanol
(inactive)	Liqua-Chrom	4, 27	44–53	10	3	Silica glass; higher capacity than others
	Zipax	10	25–37	1	1–1.5	Inactive surface; pre-coated packings available
Other	Pellidon	1, 13, 20, 27, 29	45 (av.)	1	n.a.	Nylon-bonded on glass bead; heat before packing
	Perisorb-PA6	11, 27	30–40	0.6	n.a.	Polycaprolactam layer, 2 μm thick
	Zipax-ANH	10	37–44	1	n.a.	1% cyanoethylsilicone polymer coated on Zipax

Zipax-HCP	10	25–37	1		n.a.	Non-polar saturated hydrocarbon polymer coated on Zipax; avoid temperature $> 50°C$
Zipax-PAM	10	25–37	1		n.a.	Nylon coated on Zipax
Pellumina HS	1, 5, 13, 20, 27, 29	37–44	HS 4	1		HC has a thicker coating than HS
Pellumina HC			HC 8			
Alumina						

Footnotes

1. All spherical except Actichrom and Liqua-Chrom.
2. All can be used for LLC but only the active silica and active alumina for LSC
3. n.a. means not applicable.

Suppliers

1. Alltech Associates (USA)
2. Altex Scientific (USA)
3. Analabs (USA)
4. Applied Research Laboratories (UK)
5. Applied Science Laboratories (USA)
6. Beckman Instruments (USA)
7. BioRad Laboratories (USA)
8. Corning (USA)
9. Durrum Chemical Corp (USA)
10. E. I. Du Pont De Nemours (USA)
11. E. Merck, EM Laboratories (Germany)
12. Hamilton Co. (USA)
13. Hewlett Packard (USA)
14. Hitachi (Japan)
15. J. A Jobling (UK)
16. Macherey-Nagel & Co. (Germany)
17. Perkin-Elmer (USA)
18. Phase Separations (UK)
19. Pye Unicam (UK)
20. Reeve Angel (Whatman) UK)
21. Rhone Poulenc (France)
22. Separations Group (USA)
23. Shimadzu (Japan)
24. Showa Denko KK (Japan)
25. Siemens (Germany)
26. Spectra Physics (USA)
27. Touzart and Matignon (France)
28. Toyo Soda Co. (Japan)
29. Tracor (Chromatec) (USA)
30. Varian Associates, Aerograph (USA)
31. Waters Associates (USA)
32. Shandon Southern (UK)
33. ES Industries
34. RSL Belgium
35. Chrompack (Holland)
36. Supelco (USA)
37. Regis (USA)
38. HPLC Technology (UK)
39. Laboratory Instruments Work, Prague (Czechoslovakia)
40. Brownlee Labs (USA)
41. Jones Chromatography
42. Micromeritics
43. Technicon
44. Jasco
45. Benson
46. Dionex
47. Synchron
48. MCB

when originally introduced were in the range 20–40 μm and gave poorer column efficiency than pellicular bead packings [1]. More recently the microporous particles that have been manufactured (diameters 3-10 μm) still have large surface areas (200–300 m^2 g^{-1}) but their small particle size gives more efficient columns than pellicular bead packings. This has resulted in a greater number of commercial microporous particulate packings becoming available; in 1971 the number of pellicular bead packings was 9 against zero for microporous packings, whereas in 1974 pellicular bead packings numbered 23 against 30. By 1977, there were at least 70 microporous particle packings and by 1980 few manufacturers were offering the pellicular beads except for use in precolumns or in guard columns to protect the expensive HPLC column packing. Microporous packings can also be prepared from silica, alumina, ion-exchange materials, or chemically bonded phases (Table 4.2).

Microporous particles (5–10 μm) give columns that are ten times more efficient than pellicular bead packings (40 μm); for example column efficiences of 10 000 plates per 25 cm have been achieved easily [3]. The pressure drop across a microporous packing is higher than across a pellicular bead packing because [4]:

$$\text{pressure drop} \propto 1/(d_p)^{1.8}$$

However, it has to be noted that the greater efficiency of porous particles results in shorter columns being needed, with the result that microporus particles provide faster separations using shorter columns and lower pressures.

In one respect microporous particles are more troublesome than pellicular bead packings; this is in the packing of the column. Pellicular bead packings are easy to pack dry, using the tap-fill method [5]. Microporous particles agglomerate when packed dry owing to electrostatic charges, and it has been found that a high pressure balanced-density slurry procedure is suitable.

4.3   Column Packing Techniques
The essential requirement of all column packing techniques is to produce a homogeneous bed which gives high efficiency separations. Recent work has shown that spherical particles (20–37 μm) are the smallest that can be packed dry and still give efficient columns [6]. With the arrival of smaller particle adsorbents, alternative methods have been developed using slurry packing [7].

Prior to packing the column must be cleaned. The steps required are: (a) removal of grease and particles by passage of hydrochloric acid (2 mol dm^{-3}), then chloroform; (b) drying by flow of nitrogen; (c) passage of detergent solution absorbed on cloth which is pulled through the column with a nylon fishing line; (d) removal of the detergent by passing water through the column; (e) finally, removal of the water by passage of methanol which is blown off with nitrogen.

Table 4.2 *Microporous particle packings and pre-packed columns for liquid–solid and liquid–liquid chromatography* [1a]

Type	Name	Suppliers	Average particle size (μm)	Surface area ($m^2\ g^{-1}$)	Description
Silica	BioSil A	2, 7, 27	2–10	400	Extracted with methanol and activated
Irregular	BioSil HP	7	10	350	HP-10 is 350 $m^2\ g^{-1}$ and 100 Å pore diameter HP-6 is 400 $m^2\ g^{-1}$ and 60 Å pore diameter
	Chromegasorb Si	6	5, 10	500	60 Å pore diameter, Si 100 available–100 Å pore size, filled with LiChrosorb
	Chromosorb LC-6	1, 36	5, 10	400+	Pore diameter is 120 Å, density 0.40 g cm^{-3}
	ChromaSep SL	29	5, 10, 20	400	Contains LiChrosorb Si 60
	Hi Eff MicroPart	5	5, 10	250	Pre-tested columns
	Hitachi Gel 3030 Series	14	5–7	500	pH 2–8
	LiChrosorb Si 60	2, 4, 11, 13, 25, 27, 29, 30	5, 10, 20	500	Formerly Merckosorb; Si 100 available
	MicroPak Si	30	5, 10	400	Pre-tested columns; pore size 60 Å
	Polygosil 60	16, 34, 35	5, 7, 10, 15	500	60 Å pore diameter, 0.75 ml g^{-1} pore volume
	Partisil	1, 4, 15, 19, 20, 27	5, 10, 20	400+	Columns pre-tested; size distribution given
	μPorasil	31	8–12	400	Drilled columns; pre-tested
	Silica-A	17	13±5	400	Acid washed; recommended for prep.

Table 4.2 (cont.)

Type	Name	Suppliers	Average particle size (μm)	Surface area (m² g⁻¹)	Description
	Sil-X-I	17	13±5	400	Chemically treated surface
	Zorbax SIL	10	5	300	Pre-tested columns; pore vol. 0.8 cm³ g⁻¹; 40 Å pore size
Silica Spherical	HiChrom Si	37	5	220	80 Å pore diameter, range 55–110 Å
	Hypersil	38	5–7	200	100 Å pore diameter
	LiChrospher Si 100	11, 27	5, 10	400	100 Å av. pore diameter
	Nucleosil	1, 16, 35, 38	5, 7·5, 10	300, 500	Available in Si 50 (50 Å pore diameter) with 0.8 ml g⁻¹ pore volume and Si 100 (100 Å pore diameter) with 1.0 ml g⁻¹ pore volume with 300 m² g⁻¹ surface area
	Radial-Pak Si	31	10	200	Used in Model RCM-100 radial compression module, polyethylene tube
	Separon Si VSK	39	5, 7·5, 10	450	Mean pore diameter 130 Å, 1.5 ml g⁻¹ pore volume
	Spherosil XOA 600	3, 21, 36	5–7	600±10%	80 Å average pore diameter, pore volume 0.7–1 ml g⁻¹; XOA 800 is 860 m² g⁻¹ 40 Å. 0.40–0.60 ml g⁻¹
	Spherisorb SW	1, 18, 37, 38	3, 5, 10	220	80 Å pore size, range 55–100 Å. Excaliber packing density 0.6 ml g⁻¹
	Supelcosil LC-Si	36	5	170	80 Å pore diameter, range 35–110 Å
	Spheri-5 Silica	40	5	220	

Table 4.2 (cont.)

Ultrasphere-Si	2	5		Ultrosil is 10 μm irregular
Vydac 101 TP	1, 3, 5, 22, 35	10	100	Pore diameter 330 Å, also available in irregular with 350 m² g⁻¹ and 100 Å pore diameter
Zorbax SIL	1, 10, 36, 38	5	300	Pore diameter 60–80 Å, bulk material is 8 μm
Alumina, Irregular				
Alox 60-D	16, 35	5, 10	60	60 Å pore diameter, basic, pH 9.5 activity
ChromaSep PAA	29	5, 10	70–90	Contains LiChrosorb Alox T
LiChrosorb Alóx T	2, 27, 29, 30	5, 10	70–90	Formerly Merckosorb; basic Pore diameter 150 Å
MicroPak Al	30	5, 10	70–90	Pre-tested columns
Hi Eff MicroPart Al	5	5, 10	–	Pre-tested columns
Alumina, Spherical				
Spherisorb-Al	21, 27	5, 10, 20	95	Developed by AERE Harwell, UK; dry packing procedure recommended. 130 Å pore diameter, pH limit 10; Excaliber (5), packing density 0.9 g ml⁻¹

4.3.1 Dry Packing

The packing is added in small quantities to the column which is held vertically and tapped on the bench. The sides of the column are also tapped at the level of the packing. Too much vibration may be detrimental and cause segregation of particle sizes. The procedure is as reported by Bakalyar et al. [8] and Snyder and Kirkland [9].

1. Add small portions of the packing to the column through a funnel, sufficient to fill 3–6 mm of the columm; the amount can be calculated from Table 4.3.
2. Tap the column outlet on the bench, twice a second, approximately 100 times.
3. Rotate the column and gently tap the sides just above the level of the packing during (2).
4. Gently tap the column on the floor for 15 seconds.
5. Add another portion of packing and repeat (2)–(4).
6. When the column has been filled, continue gentle tapping for about 3 minutes.
7. Clean the end of the tubing and attach a filter.

Table 4.3   *Weight of packing needed for a column 25 cm long*

Column inside diameter (mm)	Weight (g)		
	pellicular packing	*silica microporous particle*	*alumina microporous particle*
2.1	1.4	0.5	0.9
3.1	3.1	1.0	1.8
4.0	5.1	1.7	3.1
7.8	19.3	6.3	11.5

Automatic dry packers are commercially available as shown in Fig. 4.2.

4.3.2 Slurry Packing

Small particles, during dry packing, tend to form agglomerates producing a wide particle size range. Small particles, having high surface energies relative to their mass, agglomerate most severely at diameters less than 30 μm. This difficulty with small particles during dry packing has led to attempts to use solvents to aid packing. The first step in the use of solvents, or the balanced density technique, depends on the choice of two solvents, e.g. diiodomethane and tetrachloro-ethylene, that will give a mixture whose density keeps the particles suspended and prevents them from segregating out (Table 4.4). When the mixed solvents and particles are agitated vigorously, a slurry is produced.

There is a tendency for the particles to segregate out as the slurry is being transferred to the column. The relative settling rates of particles for the same column length are given: 50 μm particle diameter (1), 40 μm (1.6), 30 μm (3.0), 20 μm (6.0), 10 μm (25.0), and 5 μm (100.0); i.e. 10 μm particles settle four times as fast as 5 μm particles.

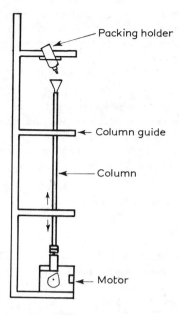

Fig. 4.2 Automatic dry column packer [8].

Table 4.4 *Properties of slurry packing solvents*

	Density (g cm^{-3})	Viscosity (cP) *at 20°C*
n-Heptane	0.7	0.4
Cyclohexane	0.8	1.0
Methanol	0.8	0.6
Ethanol	0.8	1.2
n-Propanol	0.8	2.3
n-Butanol	0.8	3.0
Dichloromethane	1.3	0.4
Bromoethane	1.5	0.4
Trichloroethylene	1.5	0.6
Chloroform	1.5	0.6
Carbon tetrachloride	1.6	1.0
Tetrachloroethylene	1.6	0.9
Iodomethane	2.3	0.5
Dibromomethane	2.5	1.0
Diiodomethane	3.3	2.9

The main variables in slurry packing are the density and viscosity of the solvent, and the flow rate and pressure of the slurry. It has been found best to use high pressures to force the slurry through the column. The use of a constant volume (metering) pump results in an increase in pressure as the column fills from a relatively low value when the column is empty. Alternatively a constant pressure pump may be used when the flow rate falls from a high value at the beginning of the operation. Constant pressure pumps seem to be the most popular because of their cost.

The steps in slurry packing for a 3.1 mm i.d. x 250 mm column are as follows [8]:

1. the solvent (250 cm^3) is degassed by heating and stirring; trichloroethylene –ethanol (1 : 1) is used for microporous silica whilst isooctane–chloroform (1 : 3) suffices for octadecyl bonded silica.

2. The packing (15% more than is needed to fill the column; see Table 4.3) is placed in a 15 cm^3 glass vial.

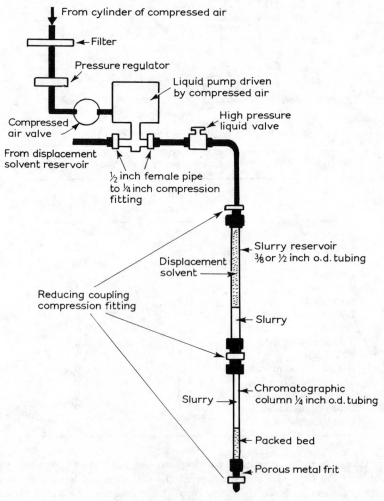

Fig. 4.3 Slurry packing apparatus [8].

3. Solvent (8 cm^3) is added to the packing and the vial is placed in an ultrasonic bath for 10 minutes.
4. The column is attached to the reservoir, and degassed solvent (2 cm^3) is added through the reservoir (Fig. 4.3).
5. The slurry is mixed throughly and poured into the reservoir.
6. Pressure (up to 6000 lbf in^{-2}) is applied to the reservoir for 5 minutes.
7. The pump is shut off, and when the flow of solvent has stopped the column can be disconnected.

In addition to balanced density solvents, such as tetrabromoethane, tetrachloroethylene, and carbon tetrachloride, ammonia stabilized slurries have been used for packing small particles [10]. Balanced viscosity systems have been recommended with solvents such as cyclohexanol and polyethylene glycol 200 [11]. 'Non-balanced density' solvents e.g. chloroform: methanol, are being used much more widely but the packing has to be done even more quickly than for 'balanced density' solvents [11, 12].

Pressure bombs with stirrers to keep the slurry in suspension are said to give a better packed column with improved efficiency.

Microporous resins based on silica and on polystyrene divinyl benzene should be 'steeped' overnight before packing into the column. If possible the packing should be done in the buffer solution in which chromatography will be performed.

4.4 Liquid–Solid Chromatography

Whilst both microporous and pellicular bead silica and alumina can be used for LSC, i.e. adsorption, the earliest pellicular bead supports such as Zipax had a relatively inactive silica bonded to a glass surface and were mainly used for liquid–liquid chromatography by adding a stationary phase. From Table 4.1 it can be seen that the active silica and alumina pellicular bead materials can be used for LSC and that most are spherical in shape. Microporous particle column materials (Table 4.2) can be either irregular in shape or spherical. Theoretically irregular particles should give higher efficiencies but spherical materials pack together better and manufacturers claim that they do not settle so readily. The microporous particles can be purchased either in bulk or ready packed in columns. The suppliers/manufacturers for each of these packings are given in the tables. It can also be seen that pore sizes range from 40 Å for Zorbax to 100 Å for LiChrospher.

In LSC, solvent molecules in the mobile phase compete with the solute molecules for sites in the solid support. Compounds with polar substituents adsorb more strongly on to the adsorbent, so LSC is best for separation of a mixture into functional group classes [13]. LSC is, in general, better for the separation of isomers because the rigid structure of the solid adsorbent surface permits interactions with one molecular shape rather than another depending on the shape.

4.5 Liquid–Liquid Chromatography

In liquid–liquid chromatography, the solute molecules partition between the mobile liquid phase and the stationary liquid phase. Differing distribution coefficients for solute molecules in the two liquids permit separation. The correct choice of the two phases means that there is an additional parameter with which to modify α in Section 2.10, providing considerable versatility for the technique. In addition to the packing methods in Section 4.3, it is also possible to add the liquid stationary phase while the support is actually in the column (*in situ* loading).

4.5.1   Mechancially Held Stationary Phases

When HPLC was first applied to LLC the solid adsorbents were coated with 10–20% of stationary phase by the solvent evaporation method used in gas chromatography. These high percentage loading phases on porous supports are difficult to pack and produced mass transfer problems resulting in band broadening. Pellicular bead materials used with 1% stationary phase such as β,β'-oxydipropionitrile or polyethylene glycol gave more easily packed, more efficient, and more homogeneous beds than large porous particle material. The newer microporous particle silica can also be used with added stationary phases.

In addition to the pellicular bead silicas in Table 4.1, LLC can be performed with nylon bonded to glass beads (Pellidons), cyanoethylsilicone polymer layer (Zipax ANH), saturated hydrocarbon polymer layer (Zipax HCP), and nylon coated silica (Zipax PAM). Perisorb-PA6 is a phase prepared from polycaprolactam layers which is valuable in that it can be used with aqueous mobile phases.

Liquid–liquid chromatography can also be performed on the microporous particles (Table 4.2) but since packing of these small particle materials with added liquid phase is difficult it has been necessary to utilize *in situ* coating methods. Low viscosity stationary phases can be loaded by pumping them through the dry packed bed, or they can be injected with a syringe into a column allowing a presaturated mobile phase to distribute the coating [14]. High viscosity phases have to be dissolved in a suitable solvent and pumped through the column, after which the solvent is removed by either a purge gas or a second solvent that precipitates the mobile phase.

These mechanically held liquid packings give satisfactory results for LLC but they suffer from a number of disadvantages. The mobile phase must be saturated with the stationary phase since otherwise the packing would gradually lose its stationary phase. This is achieved by adding a pre-column containing a high proportion of the mobile phase on wide mesh silica gel. Even when the mobile phase is saturated with respect to the stationary phase, and thus will not dissolve any of the stationary phase, stationary phase can be lost by the sheer force of the fast movement of the mobile phase. These two factors cause a gradual decrease in retention times and poorer resolution. The presence of the mobile phase in the eluent from the column can cause problems with some detectors, e.g. the differential refractometer. Finally, because the mobile phase must be saturated with the stationary phase at all times, it is not possible to use gradient elution. Many separations, unachievable previously, became possible with mechanically held stationary phases though their use is rarely reported in 1981.

4.5.2 Bonded Stationary Phases

The disadvantages of the mechanically held stationary phases in LLC prompted several workers to consider ways of chemically bonding the stationary phase to the support material. The surface of silica, which is the most popular support material, can be modified in one of three ways

1. Silicate ester formation (Si—O—R) by reaction of the surface silanol (Si—OH) groups with alcohols:

$$Si—OH + ROH \longrightarrow Si—O—R$$

2. Formation of silicone linkages (Si—O—SiR$_3$) by reaction of the surface silanol groups with an organochlorosilane:

$$Si—OH + R_3SiCl \longrightarrow Si—O—SiR_3$$

3. Formation of silicon—carbon linkage by treatment of the surface silanol groups with thionyl chloride to yield chloride which will react with organolithium compounds to produce an organic group bonded directly to the surface silica:

$$Si—OH + SOCl_2 \longrightarrow Si—Cl$$

$$Si—Cl + RLi \longrightarrow Si—R$$

The first chemically bonded packing materials were prepared by Halasz and Sebastian [15] by treating silicon chloride with 3-hydroxypropionitrile, HO(CH$_2$)$_2$CN. The bonded silicate esters had the polar cyano groups pointing away from the silica surface like the bristles of a brush and came to be known as Halasz brushes. They are commercially available as Durapak supports but suffer the known drawback of silicate esters in that they are hydrolytically unstable compared with the Si—O—Si and Si—C linked phases; i.e. they can be used only with non-aqueous solvents. However, they increase the accessibility of the stationary phase to solute molecules, effectively decreasing C_S in the van Deemter equation [Equation (2.42)]. The permanently bonded materials depending on Si—O—Si and Si—C bonds can be prepared [16] by partial hydrolysis of alkoxysilanes, followed by partial polymerization of the silanols formed and bonding of the polymer to the support. Complete cross-linking is achieved by a final heat treatment. Aue's [17] technique involves hydrolysis of silica surfaces to give additional hydroxy groups which are reacted with methyltrichlorosilane:

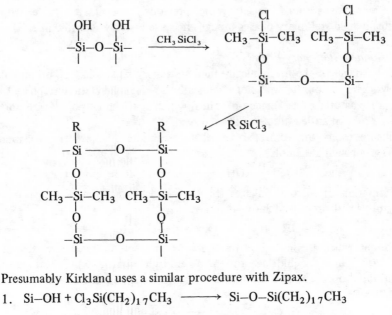

Presumably Kirkland uses a similar procedure with Zipax.

1. $Si-OH + Cl_3Si(CH_2)_{17}CH_3 \longrightarrow Si-O-Si(CH_2)_{17}CH_3$

   Zipax + octadecyltrichlorosilane $\longrightarrow$ ODS Permaphase

2. $Si-OH + (CH_3O)_3Si(CH_2)_3OCH_2CH\overset{O}{\overbrace{\quad\quad}}CH_2 \longrightarrow$

   $Si-O-Si(CH_2)_3OCH_2[CH(OH)CH_2O]_nH$

   Zipax + γ-glycidoxypropylsilane $\longrightarrow$ ETH Permaphase

3. $Si-OH + (CH_3O)_3SiCH_2CH_2CN \longrightarrow Si-O-SiCH_2CH_2CN$

   Zipax + β-cyanoethyltrimethoxysilane $\longrightarrow$ Nitrile Permaphase

These polymers cannot be removed from the surfaces by solvents, and so are permanently bonded. Microporous particles can also be used for chemical bonding as shown in Table 4.5. Other functional groups which may be added to silica include a C_8 alkyl side-chain ($SiC_7H_{14}CH_3$), phenyl ($SiPh_2$), alkylphenyl ($SiC_6H_4CH_2CH=CH_2$), fluoro-ether, and alkylamines [$Si(CH_2)_nNH_2$, which can act as an ion exchanger]. These bonded phase packings have replaced most applications of LLC with mechanically held stationary phases and are the most popular phases in use in 1981.

In Table 4.6, pellicular bead packings with bonded phase are listed. The Durapaks differ from the others in that they are silicate esters and should *not* be used with aqueous or alcoholic solvents.

The non-polar pellicular beads with their long hydrocarbon chains are especially useful for the reverse phase mode, i.e. the stationary phase is non-polar and the mobile phase is polar. Samples that are insoluble in water but soluble in alcohol or other water-miscible organic solvents, e.g. dioxan,

Table 4.5 *Microporous particle packings and pre-packed columns for bonded phase chromatography*

Type (based on functional group)	Name	Suppliers	Functionality	Base material	Particle size (μm)	Coverage	Description
Non-polar (reverse phase)	Apex ODS	4	octadecylsilane	Apex Silica	5	10%C	Endcapped
	Bio-Sil ODS-10	7	octadecylsilane	Bio-Sil HP-10	10	15%C	Medium loading
	μBondapak C_{18}	1	octadecylsilane	μPorasil	10	10%C	9000 pl m^{-1} at v = 0.44 cm g^{-1}: loading limit 20–60 mg column^{-1}
	Chromegabond MC-18	42	octadecylsilane	LiChrosorb Si 60	10	20%C	Monolayer, capped; polymeric C-18, also available with 17–18%C
	Chromosorb LC-7	1, 36	octadecylsilane	Chromosorb LC-6	3, 5, 10	15%C	Monolayer, average pore diameter 100 Å, surface area 250 m^2 g^{-1}
	HiChrom ODS	37	octadecylsilane	Spherisorb	5	0.3 mmole g^{-1}	Endcapped, pore diameter 80 Å, reversible column
	Hitachi Gel 3050 Series	14	octadecylsilane	Hitachi Gel	5–7 10–15	–	pH range 2–8
	Hypersil ODS	38, 41	octadecylsilane	Hypersil	3, 5	9%C	Monolayer, capped
	LiChrosorb RP-18	1, 5, 11, 13, 40	octadecylsilane	LiChrosorb Si	5, 10	22%C	Hibar (Merck), pH 1–9
	MicroPak MCH	30	octadecylsilane	LiChrosorb Si 60	5, 10	12%C	Monomeric layer: available in endcapped and non-endcapped versions
	MicroPak CH	30	octadecylsilane	LiChrosorb Si 60	10	22%C	Polymeric, endcapped, recommended for nonpolar molecules

Table 4.5 (cont.)

Type (based on functional group)	Name	Suppliers	Functionality	Base material	Particle size (μm)	Coverage	Description
Non-polar	Microsil C_{18}	42	octadecylsilane	Microsil	7.5	18%C	Monolayer, endcapped, surface area > 250 m^2 g^{-1}
	Nucleosil C_{18}	1, 6, 35, 38	octadecylsilane	Nucleosil 100	5, 7.5, 10	15–16%C	Capacity factor double C-8, pH 1–9, spherical
	ODS-Sil-X-1	17	octadecylsilane	Sil-X-1	13±5	–	50 cm × 3 mm columns
	Partisil ODS-1	1, 20	octadecylsilane	Partisil	5, 10	5%C	For more polar solutes, high silanol content
	Partisil-10 ODS-2	1, 20	octadecylsilane	Partisil	10	15%C	High retention and loading capacity, temperature to 70°C
	Partisil-10 ODS-3	1, 20	octadecylsilane	Partisil	10	10%C	Endcapped, 27,000 pl m^{-1}
	Polygosil C_{18}	1, 16, 35, 38	octadecylsilane	Polygosil	5, 7.5, 10	11%C	Capacity factor double C-8, pH 1–9, irregular
	Radial-Pak C_{18}	31	octadecylsilane	Radial-Pak Si	10	–	Used in Model RCM-100 radial compression module polyethylene tubing
	Separon Si C_{18}	41	octadecylsilane	Separon	5, 10	20%C	Surface area 200 m^2 g^{-1} endcapped
	Spheri-5 RP-18	40	octadecylsilane	Spherosorb	5	7%C	Coverage 0.3 mmol g^{-1}, endcapped, pore diameter 80 Å
	Spherosil-C_{18}	3, 16, 17, 36	octadecylsilane	Spherosil	5–7	20–23%C	General reverse phase; Normaton XOA 600

Name	Ref.	Silane	Base silica	Particle size (μm)	% Carbon	Comments
Spherisorb-ODS	1, 18, 37, 38, 41	octadecylsilane	Spherisorb	5, 10	7%C	Coverage 0.3 mmol g^{-1}, endcapped, pore diameter 80 Å; Excaliber (5)
Supelcosil LC-18	36	octadecylsilane	Supelcosil LC-Si	5	11.3%C	Endcapped with HMDS
Techsphere	38	octadecylsilane	Techsil	5, 10	10%C	Irregular version called Techsil C18, un-capped and capped both available
Ultrasphere-ODS	2	octadecylsilane	Ultrasphere-Si	3, 5	12%C	Endcapped, monolayer, ion pair column available
Vydac 201 C$_{18}$	1, 3, 5, 22, 35, 36	octadecylsilane	Vydac TP	5, 10	10% per 100 m^2	pH range 1–9, 3.35 μmole m^{-2}, non-endcapped, high coverage
RSIL C$_{18}$ HL	1, 34	octadecylsilane	RSIL	5, 10	18%C	Endcapped
RSIL C$_{18}$ LL	1, 34	octadecylsilane	RSIL	5, 10	8%C	Endcapped
Zorbax ODS	1, 10, 36, 38	octadecylsilane	Zorbax SIL	6	15%C	Bulk material is 8 μm, monolayer
Alltech C$_8$	1	octylsilane	Alltech Si	10	5%C	Endcapped
Apex C$_8$	41	octylsilane	Apex Silica	5	15%C	Irregular, surface area 300 m^3 g^{-1}
Chromegabond C$_8$	33	octylsilane	LiChrosorb Si 100	5,10		
FAST-LC-8	43	octylsilane	FAST-LC-Si	3	3.5 μmole m^{-2}	Spherical, monomeric, octyldimethylallyl
Finepak SIL C$_8$	44	octylsilane	Finepak SIL	10	–	Monolayer
LiChrosorb RP-8	1, 11, 13, 38, 40	octylsilane	LiChrosorb		13–14%C	Recommended for samples of moderate polarity, Hibat (Merck)
MOS-Hypersil	38, 32	octylsilane	Hypersil	3, 5, 10		Monolayer coverage, general purpose

Table 4.5 (cont.)

Type (based on functional group)	Name	Suppliers	Functionality	Base material	Particle size (μm)	Coverage	Description
	Microsil C_8	42	octylsilane	Microsil	7.5	15%C	Monolayer coverage, endcapped, surface area > 250 m² g⁻¹
	Nucleosil C_8	1, 16, 35, 38, 34	octylsilane	Nucleosil	5, 7.5, 10	10–11%C	Irregular, general purpose, pH 1–9
	Polygosil C_8	16, 35, 38, 34	octylsilane	Polygosil 60	5, 7.5, 10	10–11%C	Spherical, general purpose, pH 1–9
	Radial Pak C_8	31	octylsilane	Radial Pak Si	10		Used in Model RCM-100 radial compression module polyethylene tubing
	RSIL-C_8-D	34	octylsilane	RSIL	5, 10	–	Further deactivated after bonding
	Spheri-5 RP-8	40	octylsilane	Spherisorb	5	–	Endcapped
	Supelcosil LC-8	36	octylsilane	Supelcosil LC-Si	5	6.6%C	Endcapped with HMDS, uses octyldemethyl-chlorosilane reactant
	Techsphere C_8	38	octylsilane	Techsphere Si	5, 10	10	Irregular shaped version called Techsil C_8
	Ultrasphere-Octyl	2	octylsilane	Ultrasphere-Si	5	6.5%C	Monolayer, endcapped
	Zorbax C_8	10, 36, 38	octylsilane	Zorbax SIL	6	15%C	Bulk material is 8 μm, monolayer
Non-polar	Apex C2	41	methylsilane	Apex Silica	5	3%C	For polar/multi-functional solutes
	Chromegabond	33	methylsilane	LiChrosorb Si 100	5, 10	8%C	Also on Si 60

Name	Refs	Silane	Base material	Particle size	Loading	Notes
LiChrosorb RP-2	1, 5, 13, 38, 40, 41	methylsilane	LiChrosorb	5, 10	–	Recommended for polar compounds Hibar (Merck)
RSIL C3	1, 34	methylsilane	RSIL	5, 10	7%C	Chloropropyltrichlorosilane reactant plus TMCS endcapping
Separon Si Cl	39	methylsilane	Separon	5, 10	–	Methylated surface, useful in exclusion, exclusion limit 50 000
Supelcosil LC-1	36	methylsilane	Supelcosil LC-Si	5	2.5%C	TMCS reactant
Techsil C2	38	methylsilane	Techsil	5, 10	10%C	Useful in ion pair
Zorbax TMS	10, 38	methylsilane	Zorbax SIL	6	–	Bulk material is 8 μm, monolayer, tetramethylsilane
Chromegabond Cyclohexyl	33	hexylsilane	LiChrosorb	10	10%C	Surface area is 300 $m^2\,g^{-1}$
HiChrom C6	37	hexylsilane	Spherisorb	5	$0.6\ mmol\,g^{-1}$	$220\ m^2\,g^{-1}$, pore diameter 80 Å, end-capped, monolayer
Spherisorb C6	1, 18, 38	hexylsilane	Spherisorb	5t2	$0.6\ mmol\,g^{-1}$	$220\ m^2\,g^{-1}$, pore diameter 80 Å, end-capped, monolayer; Excaliber (5)
Apex Phenyl	41	phenylsilane	Apex Silica	5	5%C	For amines and hydroxy compounds
μBondapak Phenyl	1, 38	phenylsilane	μPorasil	10	10%C	9000 pl m^{-1} at v = 0.53 cm s^{-1}, loading limit 20–60 mg per column, recommended for more polar samples than C_{18}

Table 4.5 (cont.)

Type (based on functional group)	Name	Suppliers	Functionality	Base material	Particle size (µm)	Coverage	Description
	Chromegabond Phenyl	33	phenylsilane	LiChrosorb	10	10%C	Also for normal phase, phenyl bonded directly to Silica, surface area 500 m² g⁻¹
	Nucleosil Phenyl	1, 16, 35	phenylsilane	Nucleosil 100	7.5	10%C	Very nonpolar compounds and fatty acids
	RSIL Phenyl	1, 34	phenylsilane	RSIL	5, 10	5%C	Recommended for non-polar compounds
	Spherisorb P	1, 18, 38	phenylsilane	Spherisorb	5±2	0.3 mmol g⁻¹	220 m² g⁻¹, pore diameter 80 Å
	Techsphere C$_{22}$	38	docosanylsilane	Techsphere	5, 10	10%C	Techsil C$_{22}$ is irregular shaped particle, end-capped and non-endcapped versions available
Weakly polar	Chromegabond Diol	33	R(OH)$_2$	LiChrosorb Si 100	10		Also available on LiChrospher Si 100, 500 and 1000 Å
	Hitachi Gel 3020	14	Ester	Methacrylate	17–23		Porous polymer, spherical
	Hitachi Gel 3011	14	Aromatic	PS-DVB	10–15		Porous polymer, spherical, available in prep. size
	LiChrosorb Diol	2, 5, 40, 41	Diol	LiChrosorb	10		For very polar compounds; Hibar (Merck)

Material	Ref.	Functional group	Base silica	Particle size	Capacity/loading	Comments
Nucleosil N(CH$_3$)$_2$	1, 2, 16, 34, 35	Trialkylamine	Nucleosil 100	5, 10	—	Also weak anion exchanger, weakly basic, separation of weak acids, phenols
Nucleosil NO$_2$	1, 2, 16, 34, 35	Nitro	Nucleosil 100	5, 10	—	Spherical, affinity for double bonds
Nucleosil-OH	1, 2, 16, 34, 35	Alcoholic OH groups	Nucleosil 100	7.5	—	Wettable with water
Polygosil-NO$_2$	1, 16	Nitro	Polygosil 60	5, 10	—	Recommended for aromatics, affinity for double bonds
Polygosil 60-D-N((CH$_3$)$_2$	1	Trialkylamine	Polygosil	5, 10	—	Can also be used as weak anion exchanger
RSIL NO$_2$	1, 34	Nitro	RSIL	5, 10	5%C	Endcapped, surface area 300 m^2 g^{-1}
Separon Si CN		Cyanoethyl	Separon Si	5, 10	—	
Polar						
Alltech CN	1	Cyano	Alltech Si	10	—	Untested column
Apex Cyano	41	-CN	Apex Silica	5	Monolayer	
μBondapak CN	1	Cyano	μPorasil	10	9% by weight	Sample loading 20–60 mg per column
Chromegabond CN	33	-CN	LiChrosorb	10	7%C	
Chromosorb LC-8	1, 36	-CN	Chromosorb LC-6	5, 10	—	350 m^2 g^{-1}, average pore diameter 110 Å, density 0.43 g ml^{-1}
CPS-Hypersil	32, 38	Cyanopropyl	Hypersil	3, 5, 10	Monolayer	pH 3–8
HiChrom CN	37	Cyanopropyl	Spherisorb	5	0.6 mmol g^{-1}	220 m^2 g^{-1}, average pore diameter 80 Å, reversible column

Table 4.5 (cont.)

Type (based on functional group)	Name	Suppliers	Functionality	Base material	Particle size (μm)	Coverage	Description
	LiChrosorb CN	5, 38, 40	Cyano	LiChrosorb	5, 10	–	Prepared by first bonded tolyltrichlorosilane followed by NBS treatment and nucleophilic displacement of halogen
	MicroPak CN	30	Cyanopropyl	LiChrosorb Si 60	10	–	Aqueous compatible version shipped in methanol
	Microsil CN	43	-CN	Microsil	10	Monolayer	surface area > 200 $m^2\,g^{-1}$
	Nucleosil CN	1, 2, 16, 34, 35	Cyano	Nucleosil 100	5, 10	6 μ eq m^{-2}	–
	Partisil-10 PAC	1, 2, 38, 41	Cyano-amino	Partisil	10	–	25 000 pl m^{-1}, heptane mobile phase, 1.0 ml min^{-1}, 2:1 amino-to-cyano ratio
	Polygosil CN	1, 16, 38	Cyano	Polygosil	5, 10	6 μ eq m^2	
	RSIL CN	1, 34	Cyanopropyl	RSIL	5, 10	5%C	
	Spherisorb-CN	1, 2, 18, 36, 38, 41	Cyanopropyl	Spherisorb	5±2	0.6 mmol g^{-1}	220 $m^2\,g^{-1}$, average pore diameter 80 Å, Spheri-5(10); Excaliber (5)
	Spheri-5 Cyano	40	Cyanopropyl	Spherisorb CN	5	–	
	Techsil Nitrile	38	-CN	Techsil Si	5, 10	–	Repacked columns available

	Ultrasphere-Cyano	2	-CN	Ultrasphere-Si	5	—	Spherical
	Vydac 501 TP	1, 3, 5, 22, 35, 36	Cyano	Vydac	10	—	—
	Zorbax-CN	10, 36, 38	-CN	Zorbax SIL	6	Monolayer	Bulk material is 8 μm
Very polar	Alltech NH$_2$	1	Amino	Alltech Si	10	—	Untested column
	Apex Amino	41	-NH$_2$	Apex Si	5	—	—
	APS-Hypersil	18, 38	Aminopropyl	Hypersil	3, 5, 10	—	Special column end fitting allows guard column to butt against analytical column
	μBondapak NH$_2$	1, 38	Amino	μPorasil	10	9% by weight	pH 2–8, 9000 pl m^{-1} at $v = 0.44$ cm g^{-1}; sample loading 20–60 mg per column
	Chromegabond Diamine	33	-NH$_2$, -NH	LiChrosorb Si 60	10	—	Weak anion exchanger, also available on Si 100
	Chromosorb LC-9	36	Amino	Chromosorb LC-6	10	—	350 m^2 g^{-1}, density 0.48 g cm^{-3}
	Finepak SIL NH	44	Aminopropyl	Finepak SIL	10	—	Monolayer
	HiChrom NH$_2$	37	Aminopropyl	Spherisorb	5	0.6 mmol g^{-1}	220 m^2 g^{-1}, pore diameter 80 Å, pH < 8, reversible column
	LiChrosorb NH$_2$	1, 2, 5, 13, 38, 41	Amino	LiChrosorb	10	—	Hibar (Merck)
	MicroPak NH$_2$	30	Aminopropyl	LiChrosorb Si 60	10	—	Available in 30 cm and 25 cm × 4 mm columns or 10 g bottles

Table 4.5 (cont.)

Type (based on functional group)	Name	Suppliers	Functionality	Base material	Particle size (µm)	Coverage	Description
	Microsil NH$_2$	42	-NH$_2$	Microsil	10	Monolayer	Monolayer, coverage surface area > 250 m^2 g^{-1}
	Nucleosil NH$_2$	1, 2, 34, 35, 38	Amino	Nucleosil 100	5, 10	—	For polar compounds, weakly basic
	Polygosil -NH$_2$	1, 16, 38	Amino	Polygosil	5, 10	—	Irregular
	RSIL NH$_2$	1, 34	Aminopropyl	RSIL	5, 10	6%	Prepared by first bonding chloropropyl-trichlorosilane, followed by NBS treatment, and nucleophilic displacement of halogen
	Separon Si NH$_2$	39	Aminopropyl	Separon Si	5, 10	—	Surface area 250 m^2 g^{-1}
	Spheri-5 Amino	40	Aminopropyl	Spherisorb NH$_2$	5	—	
	Spherisorb NH$_2$	1, 36, 38, 41	Aminopropyl	Spherisorb	5, 10	0.6 mmol g^{-1}	220 m^2 g^{-1}, pore diameter 80 Å, pH < 8; Excaliber (5)
	Techsil Amino	38	-NH$_2$	Techsil silica	5, 10	—	Repacked columns available
	Zorbax NH$_2$	36, 38	-NH$_2$	Zorbax SIL	6	monolayer	Bulk material is 8 µm

Table 4.6 *Bonded phases on pellicular bead packings* [1].

Type (based on functionality)	Name	Suppliers	Functionality	Base material	Particle size
Non-polar, PLB (reverse phase)	Bondapak C$_{18}$/Corasil	31	octadecylsilane	Corasil	37–50
	CO:PELL ODS	1, 20, 27	octadecylsilane	Pellosil	37–53
	ODS-SiL-X-II	17	octadecylsilane	SiL-X-II	35±15
	Perisorb-RP	11, 13, 27	dimethylsilane	Perisorb A	30–40
	Permaphase-ODS	10	octadecylsilane	Zipax	25–37
	Vydac SC (Reverse Phase)	2, 5, 13, 16, 22	octadecylsilane	Vydac Adsorbent	30–44
Polar, PLB	Bondapak Phenyl/Corasil	31	phenylsilane	Corasil	37–50
	CO:PELL PAC	1, 20, 27	nitrile	Pellosil	37–53
	Durapak Carbowax 400/Corasil	31	polyethylene oxide	Corasil	37–50
	Permaphase-ETH	10	ether	Zipax	25–37
	Vydac SC-Polar	2, 5, 13, 22, 27, 29, 30	ethyl nitrile	Vydac Adsorbent	30–44

All spherical shape.

acetonitrile and tetrahydrofuran, are candidates for reverse phase chromatography. These phases are all based on the pellicular bead materials exemplified in Table 4.1. Bonded phase coverage depends on the number of silanols available and on the type of reaction. Low surface area silicas (e.g. Zipax) require polymerization [18] but for high surface area silicas (e.g. LiChrosorb) direct reaction without polymerization is satisfactory [19].

The alkyl phase loading (as a % carbon by weight) can be varied from 5% as in Partisil [I] ODS to 22% in Lichrosorb RP18. When the loading is low (5-10%) there is probably a monolayer coverage whilst the higher loading (12-25%) suggests a polymeric coverage [20].

The monomeric phases ought to have a higher separation efficiency since the mass transfer rate is slower on the polymeric layer. Against this, any residual silanols which may act as adsorption sites are more likely to be blocked with the polymeric layer reducing the chance of tailing.

When di- and trichloro or alkoxysilanes are used, additional silanols are produced by the hydrolysis of the unreacted reagent functional groups. If such a bonded phase is treated with trimethylchlorosilane, many of these silanols can be blocked, and the stationary phase is said to be 'capped'.

The polar phases can be used with normal LLC mode (e.g. polar stationary phase or a non-polar mobile phase) or with reverse phase mode. These polar phases have either ether or nitrile functional groups. Bonded phases on large porous particles are given in Table 4.7 though they have little market in 1981.

Table 4.7   *Bonded phases on large porous particles* [1].

Type	Name	Functionality	Base material
Non-polar, porous	Bondapak C_{18}/Porasil	octadecylsilane	Porasil B
	Bondapak Phenyl/Porasil	phenyl	Porasil B
Polar, porous	DuraPak OPN/Porasil	nitrile	Porasil C
	DuraPak Carbowax 400/Porasil	polyethylene oxide	Porasil C

1. All spherical shape and 37–75 μm size.
2. All supplied by supplier 31.

4.5.3   *Preparation of Bonded Phases in situ*

As an alternative to the purchase of ready-made bonded phases, there have been reports of their on-column preparation [21]. Some of the steps are given here [22].

Teflon-lined stainless-steel tubing (2 ft x 1/8 inch o.d (61 x 0.32 cm)) is rinsed with methanol and chloroform. The column is dried by the passage of nitrogen and then packed with Corasil II (37–50 μm) by gentle vibration and suction.

A Teflon-lined stainless steel reservoir (6 ft x 1/8 inch o.d. (182 x 0.32 cm)) is filled with 10% octadecyltrichlorosilane in toluene and coupled to the Corasil II column. The column reservoir is attached to an LC pump which is fed from a solvent reservoir containing wet toluene. The silanizing solution is pumped through the column at a rate of 0.5 cm^3 min^{-1}. After all the silane−toluene mixture has been eluted from the column, a further portion of toluene (11 cm^3) is pumped through the column. Acetonitrile (100 cm^3) and acetonitrile−0.01 M -ammonium carbonate (80 : 20) (100 cm^3) are used to rinse out the columns which are then dried at 150°C for 12 hours with a stream of dry nitrogen.

Phenyltrichlorosilane can be used in place of octadecyltrichlorosilane. When dry toluene replaces wet toluene, the Corasil II column must be dried at 150°C for 4 hours with a flow of nitrogen gas before treatment with the silanizing solution.

4.5.4 Similarity of Commercial Packings

For simplicity, we said in Section 4.2.1 that many differently named products will do the same job. The apparently vast numbers of packings available in earlier days have been reduced to such an extent that most manufacturers will recommend:

1. a silica microporous particle column;
2. a reverse phase C_{18} bonded phase column;
3. a more polar bonded phase column e.g. CN or NH_2.

They will suggest that these three types should be satisfactory for most separations. It is important to recognize that one manufacturer's C_{18} bonded phase may not give identical results to that of another supplier. Goldberg [23] has shown in Table 4.8 that the relative retention (α) for terphenyl to biphenyl can vary from 3.40 to 0.90.

Table 4.8 *Relative retention of commercial packings.*

	$\dfrac{K' \text{ terphenyl}}{K' \text{ biphenyl}}$		$\dfrac{K' \text{ caffeine}}{K' \text{ theophylline}}$
Brownlee RP18	3.40	Zorbax ODS	2.70
Ultrasphere ODS	2.90	Partisil ODS	2.20
Zorbax ODS	2.75	Brownlee RP18	2.15
Micropak MCH 10 ODS	2.70	Partisil ODS −2	2.10
Partisil ODS	2.60	ODS Sil−X−1	1.80
ODS hypersil	2.30	Ultrasphere ODS	1.70
Bondapak C_{18}	2.10		
ODS Sil−X−1	0.90		

To judge column performance, column efficiency, peak symmetry, selectivity and permeability can be used.

The column efficiency is measured by the number of theoretical plates where the larger the number of theoretical plates the more likely the column will carry out the expected separation. Values in excess of 60 000 theoretical plates per metre are now achieved with the existing column packing technologies. The number of theoretical plates can be calculated by drawing tangents to the peaks and using Equations (2.21) and (2.22). Alternatively if the signal is processed by an integrator, the number of theoretical plates can be calculated from

$$N = 2\pi \left(\frac{h \times t_R}{A} \right)^2$$

where h is peak height, t_R is the retention time and A is the peak area.

Peak asymmetry or peak tailing can be hand calculated by using the values shown in Fig. 4.4. The peak asymmetry factor is given by $\dfrac{CB}{AC}$ which are measured at 10% of peak height. A peak asymmetry factor of 2.4 or less represents a good column. A more accurate method of determining peak tailing depends on a computer calculation using a skewed peak model of the exponentially modified Gaussian function. The peak contour is described by the convolute integral between a Gaussian constituent having the standard deviation σ and an exponential modifier having the time constant τ (Fig. 4.5).

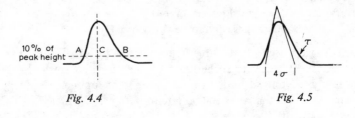

Fig. 4.4   Fig. 4.5

Peak symmetry can be measured by a computer calculated value for peak skew (Fig. 4.5).

$$\text{Peak skew} = \frac{2(\tau/\sigma)^2}{[1+(\tau/\sigma)^2]^{\frac{3}{2}}} \qquad t_w = 4\sigma$$

Where σ is the Gaussian component and τ is the exponential tailing component.

Peak tailing increases with the ratio τ/σ.

Thus τ/σ value of 0.1 and a skew value of 0.002 represents a good symmetrical peak, whereas the τ/σ value of 2.0 and a skew value of 1.43 indicates a very asymmetrical peak with a lot of tailing. It is usual to operate in the range: peak skew 0.7–1.6 with an asymmetry factor of 1.25–2.1.

A test for column to column reproducibility is the selectivity factor α which was given in equation 2.49 as the ratio of the capacity factors for two solutes.

Permeability (Equation 2.60) can be visualized by specifying a measured flow rate for a given column with a specific mobile phase at a fixed pressure and temperature. It is also possible to give the same information by giving the instrumental pressure required to produce a given flow rate with a specific mobile phase and known temperature.

Bonded phase characteristics should include the description that it is monomeric or polymeric. If the phase is polymeric, it is helpful to quote the percentage organic or carbon. If the phase is monomeric it is helpful to calculate the surface coverage as

$$\text{Coverage} = \frac{W}{M \times S}$$

where W is the weight of organic layer, M is the mol. wt. of the bonded group and S is the specific surface area of the adsorbent.

Monomeric phases tend to give higher column efficiencies.

4.6 Ion-Exchange Chromatography

Instrumental liquid chromatography was developed from the amino acid analyser technique of Moore *et al.* [24] which uses the ion-exchange mode of liquid chromatography first employed in the 1850s using zeolites. The column packing material is a solid porous matrix with fixed ionogenic groups — usually an ion-exchange resin. Most conventional ion-exchange resins are cross-linked organic polymers forming a solid porous matrix to which are attached the ionogenic groups. The two types of ion exchangers, cation and anion, can be further sub-divided into strong and weak depending on the dissociation constant of the ionogenic groups of the resin (Table 4.9).

Conventional resin beads are prepared by copolymerization of styrene and divinylbenzene with a degree of cross-linking to provide mechanical stability. They may be of two types, microreticular (gel) and macroreticular (macroporous). The gel type becomes porous on swelling, with the pore size dependent on the degree of cross-linking indicated by the percentage of divinylbenzene in the copolymer. In addition to the polyaromatic polymers, cellulose and dextran act as suitable polymers from which ionogenic groups can be suspended, e.g. diethylaminoethyl-(DEAE) and carboxymethyl-cellulose.

Table 4.9 *Ionogenic groups on ion-exchange resins*

Class	Ionogenic group
Strong cation exchanger SCX	$-SO_3H$
Weak cation exchanger WCX	$-CO_2H$
Strong anion exchanger SAX	$-\overset{+}{N}R_3$
Weak anion exchanger WAX	$-NH_2$

From these original gel type ion exchangers, porous and pellicular bead materials have been developed for HPLC. Pellicular bead exchangers have been produced from a solid polymer core which has ion-exchange resin only on the surface, or from a glass bead with a skin of ion-exchange material. The conventional resins can also be modified by using a porous surface with ion-exchange properties [13]. New pellicular materials have been developed using pellicular silica which has been treated with organosilanes containing ionogenic groups, e.g. $Corasil-Si-O-Si(CH_2)_3 N^+(CH_3)_3$.

Although pellicular bead material is not used for many applications in liquid–liquid and liquid–solid chromatography, it has retained its popularity for ion-exchange chromatography. This is because the pellicular beads can give significant

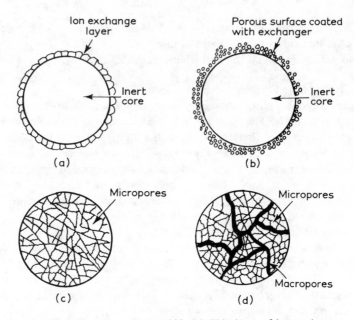

Fig. 4.6 Ion-exchange packings [13]. (a) Thin layer of ion exchanger on solid inert core. (b) Porous surface coated with exchanger on an inert core. (c) Microreticular ion exchanger with small pores. (d) Macroreticular ion exchanger with macropores.

differences in retention behaviour compared with porous resins or silica-based exchangers. They show lower pressure drops with aqueous solutions though their larger particle sizes cause lower efficiencies. They have lower ion-exchange capacity so that the buffer solutions need to be of lower strength. The larger particle size allows them to be dry-packed.

Ion-exchange chromatography depends on the fact that the positive charge of quaternary ammonium ion will attract and exchange anions, i.e. it is an anion exchanger. As a buffer solution of negatively charged counter-ions is allowed to flow over the ion exchanger, the positive charge on its surface attracts the counter-ions:

$$\text{Corasil—Si—O—Si(CH}_2)_3\overset{+}{\text{N}}\text{(CH}_3)_3 \xrightleftharpoons{\text{Na}^+\ ^-\text{OCOCH}_3}$$

$$\text{Corasil—Si—O—Si(CH}_2)_3\overset{+}{\text{N}}\text{(CH}_3)_3\ \overline{\text{O}}\text{COCH}_3$$

If a mixture of benzoic and *o*-toluic acids (pK_a 4.2 and 3.9 respectively) is passed over the ion exchanger at a pH of 5.7 where they are fully ionized, their anions will exchange for the acetate ions on the surface:

$$\text{Corasil}-\text{Si}-\text{O}-\text{Si}(\text{CH}_2)_3\overset{+}{\text{N}}(\text{CH}_3)_3 {}^-\text{OCOCH}_3 \quad \xrightleftharpoons{\quad \text{PhCOO}^- \quad}$$
$$\text{Corasil}-\text{Si}-\text{O}-\text{Si}(\text{CH}_2)_3\overset{+}{\text{N}}(\text{CH}_3)_3 \; \bar{\text{O}}\text{COPh}$$

All strong exchangers are dissociated in the pH range 1–13; weak exchangers tend to have a pK_b around 6.5, so the pH of the buffer solution has to be below 5. The acid anions are held on the ion-exchange surface by Coulombic forces and can be displaced by changing the pH of the buffer solution used as eluent. If the buffer solution is started at a pH above the pK_a of the acids and changed to convert the acids back into neutral molecules, the acids will be eluted in the order of decreasing pK_a.

The capacity of an exchanger is the number of ionogenic groups per unit weight or volume of exchanger. For polystyrene based conventional exchangers the usual capacities are of the order of 3 and 5 milliequivalents per gram of dry resin for anion and cation exchangers respectively. To ensure that the efficiency of the column is not reduced by overloading, the sample size is normally kept to less than 5% of the ion-exchange capacity. The capacity is, however, a function of the pH. The ionization of a cation exchanger R_cH to produce the resin ion $R_c{}^-$ and hydrogen ions H^+ will be suppressed by an excess of H^+ ions, i.e. at low pH. Similarly an anion exchanger, ionizing to give OH^- ions, will have its ionization suppressed by a high concentration of OH^- ions, i.e. at high pH. This sets limits on the pH range in which the exchanger can be used. Thus strongly and weakly acidic cation exchangers can be used in the pH range 2–14 and 8–14 respectively, whereas strongly and weakly basic anion exchangers have useful pH range 2–10 and 2–6 respectively.

Most of the pellicular ion exchangers (Table 4.10) have a low ion-exchange capacity ($5-15\ \mu\text{equiv g}^{-1}$) but some with higher capacity ($100\ \mu\text{equiv g}^{-1}$) can be used for large sample sizes. The earlier resins tended to swell and shrink with changes in pH, ionic strength, and temperature even though the glass bead was unaffected and the pellicular resins tended to flake. Many newer packings (Bondapak CX/Corasil, Permaphase AAX, Perisorb PA) consist of silica pellicular beads that have been treated with organosilanes having either positively or negatively charged groups. They can be used with non-aqueous modifier without loss of stationary phase. Pellicular bead exchangers with their low capacity are easily poisoned. The pores in conventional resins cause slow mass transfer and some size separation occurs. The fast mass transfer in pellicular exchangers permits analysis time to be reduced by factors of 10.

One of the advantages of pellicular materials over conventional resins is that they can be packed readily by the dry packing procedure to give highly reproducible packings. They do not swell in water and are stable towards changes in eluent composition, both of which lead to a long life. The biggest disadvantage of the pellicular resins involves the relatively low phase ratio producing low column loading capacity so that trace analysis is exceptionally

Table 4.10 *Pellicular ion-exchange resins* [1].

Type	Name	Suppliers	Particle size (μm)	Base*	Ion-exchange capacity (dry) (μequiv g^{-1})	Description
Anion	AE-Pellionex-SAX	1, 20, 27, 29	44–53	PS–AE		$T = 70°C$; pH 2–10; can be used with organic solvents
	AS-Pellionex-SAX	1, 20, 27, 29, 30	44–53	PS–DVB	10	May be used up to 85°C; pH 2–12; can be used with organic solvents
	AL-Pellionex-WAX	1, 20, 27, 29	44–53	AL	–	pH 2–7; $T = 75°C$
	Bondapak/AX/Corasil	31	37–50	BP	10	pH 2–7; can be used with organic solvents
	Perisorb-AN	11, 13, 27	30–40	BP	30	pH 1–9
	Permaphase-AAX, Permaphase-ABX	10	37–44	BP	10	Stable to 75°C; pH 2–9; can be used with organic solvents; avoid strong oxidizing agents; 1% stationary phase
	Vydac SC Anion	5, 16, 22, 27, 29	30–44	PS–DVB	100	Supplied in chloride form; can be used with organic solvents

	Zipax-SAX	10	25–37		12	Lauryl methacrylate polymer base; M.W. 500; pH 4–10
	Zipax-WAX	10	25–37	PAM	12	Polyamide base
Cation	Bondapak CX/Corasil	31	37–50	BP	30–40	pH 2–8; $T = 60°C$
	HC- or HS-Pellionex-SCX	1, 20, 27, 29, 30	37–53	PS–DVB	HS 8–10 HC 60	pH 2–10; $T = 40°C$; can be used with organic solvents
	Perisorb-KAT	11, 13, 27	30–40	BP	50	pH 1–9
	Zipax SCX	10	10	FC	5	Only small amounts of organic solvents; fluoro-polymer base, M.W. 1000
	Vydac SC Cation	5, 13, 16, 22, 27, 29	30–44	PS–DVB	100	Supplied in hydrogen form

*PS–DVB = polystyrene–divinylbenzene copolymer
PS–AE = polystyrene–aliphatic ester copolymer
FC = fluoropolymer base
BP = bond phase through siloxane base
AL = aliphatic

1. All anion exchangers have NR_3^+ except Al-Pellionex-WAX and Zipax WAX which have NH_2 functionality and are weakly basic.
2. All cation exchangers have SO_3^- and are strongly acidic.

difficult. This low phase ratio also affects the eluent composition, permitting weaker eluents to be used.

Porous resins (Table 4.11) have been prepared as small particles (< 25 μm) which show good column efficiencies and high exchange capacities though they are less efficient than silica micro-particles of the same size owing to slower intraparticle diffusion (stationary phase mass transfer). All have ion exchange capacity (1-2 mequiv g^{-1}) much greater than that of pellicular packings. Most microporous exchangers utilitize the polystyrene–divinylbenzene resin as the base material but an increasing number are coated on silica beads. This bonding can be with organosilanes which have ionogenic groups present. Microporous ion exchangers tend to give higher column efficiencies than the pellicular packings which have large particles. Fast separations can be achieved by reducing the column length; e.g. nucleoside phosphates can be separated in 15 minutes compared with 150 minutes on a pellicular phase. Knox has shown that chemically bonded porous anion exchangers, e.g. Permaphase AAX, are significantly superior in kinetic terms to conventional ion-exchange resin beads or porous layer ion exchangers as measured by logarithmic plots of reduced plate height against reduced velocity [25].

Generally the cross-linked polystyrene–divinylbenzene cation exchanger resins are the most stable with respect to temperature, and some may be used up to 150°C. In contrast, the anion exchangers with the polystyrene–divinylbenzene matrix are stable only up to 80°C, and some only up to 40°C. The pellicular resins are even less stable and many can be used only at room temperature.

4.7 Exclusion Chromatography

Exclusion chromatography is a technique that permits solutes to be separated by their effective size in solutions, i.e. from molecular weights of 100 to 8×10^5. Elution is performed on a rigid, porous, non-ionic support whose pores are similar in size to those of the sample molecules. Conventional exclusion chromatography was first utilized to fractionate natural polymers in aqueous solution, and more recently, as solvent resistant materials have become available, to permit analysis of synthetic polymers.

Conventional exclusion chromatography support materials may be subdivided into three categories: xerogels, aerogels, and xerogel–aerogel hybrids. Xerogels consist of a solution of linear macromolecules whose movement is restricted by cross-linking or by physical interaction [26].

The first supports used in exclusion chromatography, cross-linked dextran (Sephadex) and cross-linked polyacrylamide (Biogel P), were xerogels used with aqueous media (Fig. 4.8). Additionally, Sephadex LH20, the hydroxypropyl ether of cross-linked Sephadex, and Enzacryl, cross-linked polyacryloylmorpholine, were developed to permit chromatography in organic solvents. They cannot be used in HPLC.

Aerogels consist of a rigid matrix containing pores, e.g. porous glass and

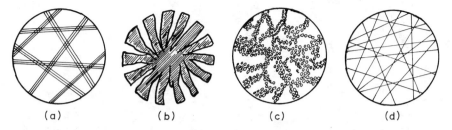

Fig. 4.7 Conventional exclusion chromatographic supports [26]. (a) Typical xerogel. (b) Aerogel (porous silica). (c) Aerogel (porous glass). (d) Xerogel-aerogel hybrid (agarose gel).

porous silica into which a solvent has been introduced. They are not really gels and do *not* actually collapse when the solvent is removed. Xerogel—aerogel hybrids have features common to both, i.e. the polymer is a semi-rigid structure which undergoes very little dissolution on forming the gel. The hybrids have very large pore sizes and a polymer matrix that is very resistant to compressive strains, e.g. agarose gels ('Biogel A, Gelarose, Sagavac and Sepharose), cross-linked polystyrene (Styragel), sulphonated macroreticular polystyrene (aquapak), and cross-linked polyvinylacetate (Merck O—Gel OR). Some of the formulae for these materials are given in Fig. 4.8.

For HPLC the usual constraints apply; small particle size results in rapid mass transfer, faster flow rates, and shorter analysis times, e.g. minutes instead of hours. The commercial packings (Table 4.12) are either of the cross-linked polystyrene type and are semi-rigid or of the controlled pore size glasses or silicas which are rigid. Most cross-linked polystyrene beads swell in solvent and cannot be used at too high a pressure, 100–150 atm. The semi-rigid materials have larger pore volumes and can separate smaller particles than the rigid particles. As aerogels, the rigid packings can be used with either lipophilic or hydrophilic solvents but the semi-rigid packings are mainly xerogel—aerogel hybrids for use with non-aqueous solvents.

Solvent selection in exclusion chromatography is easier than for other HPLC modes and only one solvent is needed to dissolve the solutes and to elute them. All solute molecules elute between the excluded volume and the total permeation volume which for microporous particle columns (100 Å pore size 10 μm in 25 X 0.6 cm) would be 2.5-3.5 cm^3. Analysis time is very short, i.e. 5–30 min.

4.8 Care of Columns

It has been recognized that the sinking of the top of column packings is not caused by excess pressure but by the dissolution of the silica. This sinking alters the level at the top of the analytical column which in turn causes the shape of the eluted peak to alter. To minimize this problem some manufacturers recommend that a precolumn of 5 μm silica be placed between the pump and the injector where it will saturate the mobile phase with respect to silica so that the solvent will not damage the analytical silica by dissolution. It has been found that 37–53

Table 4.11 *Microporous particle ion exchangers for high performance liquid chromatography*

Type	Name	Suppliers	Particle size (μm)	Ion exchange capacity (dry) (mequiv g^{-1})	Description
Anion-based on divinylbenzene-polystyrene	Aminex A-Series	2, 7, 27, 30	A-14 20±3 A-25 17.5±2 A-27 13.5±1.5 A-28 9±2	3.4 A-14 3.2 others	A-28 can withstand pressure of 10 000 lbf in^{-2}
	AN-X	35	11	4	2, 4, 8 and 12% crosslinked
	Benson BA-X	1, 45	7–10	5	pH 1–12, sugar, carbohydrates, crosslinking X4, X6, X8, X10, also for amino acids
	Benson BWA	45	7–10	5	10–15 also available
	Chromex	9, 46	11±1	4	Crosslinking of 2, 4, 8, or 12%. An 8% crosslinking available with $d_p = 8$ μm
	Durrum DA-X8F	9	11±1	2	Also DA-X4 20±5 μm; for acidic biomolecules, peptides
	Durrum DA-X8F	9	8±1	4	Also DA-X8 11±1 μm; for anions in urine
	Hamilton 7800 Series	4, 12, 27	20±5	5	For carbohydrates, nucleotides
	Hamilton-HA	5, 12	7–11	5	4, 6, 8 or 10% crosslinked
	Hitachi Gel 3011-N	14	10–15	–	pH 2–11, nucleotides, organic acids, spherical, custom resins available for sugars
	Ionex 5B	2, 16, 34	5–20	3	7% crosslinked
	Shodex Axpak	24	–	–	Packed in dilute acetic acid

Anion–Silica-based				
μBondapak-NH$_2$	31	10	–	pH 4–9
Chromegabond SAX	33	10	1	on 100 Å silica
Lichrosorb AN	1, 2, 11	10	0.55	pH 2–9
MicroPak-NH$_2$	30	10	–	For nucleotide isomers, highly phosphorylated nucleotides, general purpose anion exchange
MicroPak SAX	30	10	–	
Microsil SAX	42	10	1	Columns only
Nucleosil-NH$_2$ or -N(CH$_3$)$_2$	16	5–10	–	pH 2–9
Nucleosil-SB	16	5–10	1	pH 0–9; may be used with organic eluents
Partisil 10 SAX	2, 20, 27	10	<1 (est.)	14 000 pl m^{-1}, pH range 1.5 to 7.5, supplied in H$_2$PO$_4$ form in methanol
RSIL AN	1, 34	5, 10	<1 (est.)	7% loading
SynchropakAX-300	5, 37, 50	10	–	300 Å pore size, reccommended for proteins and enzymes, monolayer coverage, pH range 2–8
Vydac 301 TP	1, 3, 5, 22, 35	10	0.2	pH range 1–9, high density hydrophobic phase protects silica
Zorbax SAX	10, 38	7	<1 (est.)	pH 2–9

Table 4.11 (cont.)

Type	Name	Suppliers	Particle size (μm)	Ion exchange capacity (dry) (mequiv g^{-1})	Description
Cation-based on divinylbenzene-polystyrene	Aminex A-Series	2, 7, 27, 30	A-4 20±4 A-5 13±2 A-6 17.5±2 A-7 7–11	5	A-7 can withstand pressure of 10 000 lbf in^{-2}
	Aminex HPX-87	2, 7, 27, 30	9	5	H$^+$ form called ion exclusion column for organic acids, Ca$^+$ for use for carbohydrates; HPX-42 is 22 μm, Ca$^+$ form, and recommended for oligosaccharides, heavy metal for mono- and di-saccharides.
	Beckman AA-Series	6	AA-20 10.5±1 AA-15 11±3	5	Made for amino acid analysis
	Beckmann PA-Series	6	PA-35 11±3 PA-28 16±4	5	Made for amino acid analysis
	Benson BC-X	1, 45	7–10, 10–15	5.2	pH 1–14, can also be used in RPC, crosslinking x4 to x32, several other cation resins available for amino acid and peptide analysis
	Benson BCOOH	45	7–10	10	Useful for hemoglobin fractions, drugs, and proteins
	Chelex 100	7	25±5	2.9	Chelating resin; for ligand exchange; 68–76% water
	Chromex cation	9	11	5	8 or 12% crosslinked
	Dionex DC	46	−1A 14±2 −4A 9±0.5 −6A 11±1	5	Crosslinking 8%. Recommended for amino acid analysis
	Durrum-DC-1A	9	14±2	5	For protein hydrolysates, amino acid analysis

	Durrum-DC-4A	9	9±0.5	5	For amino acid analysis; single column
	Durrum-DC-6A	9	11±1	5	For amino acid analysis; single column
	Hamilton H-70	4, 12, 27	24±6	5.2	For acidic and neutral amino acid analysis; hydrolysate analysis only
	Hamilton HP-Series	4, 12, 27	AN 90 13.5±6.5 B 80 8.5±1.5	5.2	B-80 For basic amino acid analysis; AN-90 acidic and neutral amino acids
	Hamilton HC	4, 12, 27	10–15	5	Sold in bulk, Ca$^+$ form available for carbohydrates
	Hitachi Gel 3011-C	14	10–15	–	pH 2–11, separation of amines, spherical
	Hitachi Gel 3011-S	14	10–15	–	Nucleic acid bases, amino acids, speciality amino acid resins are available in 5 μm
	Ionex SA	16	10±2 15±2.5 20±3	5	Moisture 55%
	Shodex Cxpak	24	–	–	Packed in citric acid buffer, amino acids and saccharides
Cation–Silica-based	Chromegabond SCX	6	10	<1 (estimate)	60 Å pore size
	LiChrosorb-KAT	1, 5, 38, 40	10	1.2	On 100 Å silica
	Microsil SCX	42	10	1	Columns only
	Nucleosil SA	1, 2, 16, 35	5, 10	~1	pH range 1–9
	Partisil-10 SCX	1, 2, 20	10	~1 (estimate)	Supplied in NH$_4^+$, nucleic acids, vitamins, pH 1.5–7.5
	RSIL-CAT	1, 34	5, 10	–	5% loading
	Vydac 401 TP	1, 3, 22, 36	10	~1 (estimate)	Metals, amines, N-bases
	Zorbax SCX	10, 38	6–8	5	pH 2–9, nucleic acid constituents, water soluble vitamins

Table 4.12 *Small porous packings for exclusion chromatography*

Name	Suppliers	Particle size (μm)	Base material	Pore sizes (Å) or M.W. ranges (PS)	Description
Benson BN-X	1, 45	7–10	PS-DVB	1500 × 4 500 × 7	Other crosslinking available, 16 000 pl m⁻¹, can be used with aqueous mobile phases also, sold in bulk
BioBeads S	7	10	–	4×10^2	12% crosslinking
Bio gel P-2	7	28	Polyacrylamide	1.8×10^3	Too soft for HPLC
Bondagel	31	10	Silica modified with ether groups	$2 \times 10^4, 1 \times 10^5$ $5 \times 10^5, 2 \times 10^6$	Packed column only
CPG	8	5–10	Porous glass	40, 100, 250, 550, 1500 Å	Can be used in nonaqueous and aqueous to pH 8. Pore volumes vary from 0.1–1.5 ml g⁻¹. Glycophase G has glycerol coating
Chromegapore	6	10	LiChrospher (Silica)	60 100 500 1000	60 Å base material is LiChrosorb, irregular, available as deactivated and as protein columns
Chromex	2	10	PS-DVB	1000 (E-1) $500 - 2 \times 10^4$ (E-20) $10^2 - 10^5$ (E-100) $10^4 - 2 \times 10^6$ (E-2000)	Bulk only, can also be used as reverse phase packing
EM gel Type	11	30–60	Vinylacetate copolymer	1.5×10^3 8×10^3	Too soft for HPLC
EM type	11	10–40	Silica-untreated irregular particles	5×10^4 4×10^5 1×10^6	
Finepak Gel 101	44	8	PS-DVB	MW < 3000	40 000 pl m⁻¹ guaranteed

Name	Ref		Material	Molecular weight range	Comments
HSG	23	10	PS-DVB	4×10^2 4×10^3 4×10^4 4×10^5 2×10^6 8×10^6 (estimate) 10^7 (estimate)	Packed columns, 50 cm × 8 mm. Compatible only with non-aqueous solvents
Ionpak	24	10–15	PS-DVB Sulfonate	1×10^3 5×10^3 5×10^4 5×10^5 5×10^6 (estimate) 5×10^7 (estimate)	S-800 series, columns only
LiChrospher	1, 5, 38	10	Silica	100 $5–8 \times 10^4$ 500 $3–6 \times 10^5$ 1000 $0.6–1.4 \times 10^6$ 4000 $2–5.8 \times 10^6$	Spherical, pore volume of Si 100 is 1.2 ml g^{-1}; for others, 0.8 ml g^{-1}, bulk density 0.38 – 0.42 g ml^{-1}. Maximum pressure 5000 lbf in^{-2}
LiChrospher Diol	51	10	LiChrospher (silica)	100 500 1000 4000	$\begin{array}{ccc} & & OH \;\; OH \\ & & | \quad\quad | \\ —C & —C & —C— \text{ phase, LiChrosorb Diol is} \\ & & | \quad\quad | \\ & & OH \;\; OH \end{array}$ similar but on irregular silica, packed columns only
MicroPak TSK Type H	28, 30	8–10	PS-DVB	40 $0.05–1 \;(\times 10^3)$ 2.5×10^2 $0.05–6 \;(\times 10^3)$ 1.5×10^3 $1–60 \;(\times 10^3)$ 1×10^4 $0.5–40 \;(\times 10^4)$ 1×10^5 $0.2–400 \;(\times 10^4)$ 10^6 (estimate) 10^7 10^7 (estimate) 10^8 $1.5 \times 10^3–10^7$ (mixed) $10^3–10^8$	Packed columns, 30 cm and 50 cm × 8 mm, packed in THF but available in other solvents, larger pore sizes available on request. TSK 100OH–700OH
MicropakBKG	30	10	PS–DVB	$2 \times 10^3, 1 \times 10^4$ $8 \times 10^4, 4 \times 10^5$ 3×10^6	Packed column only

Table 4:12 (cont.)

Name	Suppliers	Particle size (μm)	Base material	Pore sizes (Å) or M.W. ranges (PS)	Description
Ohpak	24	15–20	Hydroxylated Polyester	5×10^4 4×10^5 2×10^6 1×10^7 (estimate)	B-800 series, columns only
PL Gel	5, 38	10	PS-DVB	50 100 500 10^3 10^4 10^5 10^6	High crosslinking, 40 000 pl m^{-1}
Porasil 60 GPC	31	10	Spherical silica	1×10^4	
Protein Column	31	10	Silica	125 2000–8000 Daltons Proteins 250 10 000–500 000 Daltons	Column dimensions 7.9 × 30 cm
RSIL-PP	1, 34	10	Silica	100 200 500	Acetylaminopropyl bonded phase, 10% coverage
Shodex	24	10	Polystyrene DVB	1.5×10^3, 5×10^3, 7×10^4, 8×10^5, 8×10^6, 5×10^7 (PS)	Solvent DMF, AD-800 series Solvent THF, A-800 series Prep column available
Spherosil	21, 27	7	Silica glass	80, 140 360, 680, 1100 Å	Rigid spherical packing; may have adsorption with polar molecules
μSpheragel	2	–	PS-DVB	50 <2000 100 100–5000 500 5000–10^4 10^3 10^3–5×10^4 10^4 10^4–5×10^5 10^5 10^5–5×10^6 10^6 <10^6	18 000 pl m^{-1}, sample loading capacity 200–500 mg per column

Name	Ref.	Particle size	Pore size / Molecular weight	Material	Comments
μStyragel	31	—	100 — 0–700; 500 — 500–10 000; 10^3 — 1000–20 000; 10^4 — 10 000–200 000; 10^5 — 100 000–2 000 000; 10^6 — 10^6–$>2 \times 10^7$	PS-DVB	12 000 pl m^{-1} for 100 Å column, 9000 pl m^{-1} for others at $v = 0.18$ cm s^{-1}. Loading limit 10 mg per column
Synchropak GPC	5, 37	10	100 — 1×10^5; 300 — 5×10^5; 500 — 2×10^6 Dextran; 1000 — 4×10^7 (estimate); 4000 — 10^8 (estimate)	LiChrospher (Silica)	300 Å base material is Vydac TP Si, silica-base, available in bulk or columns; hydrophilic carbohydrate monolayer phase, Aquapore (Brownlee and Chromatix)
TSK Gel G	28, 30	10	8×10^4; 2×10^6	Hydroxylated Organic	Packed column only
TSK Type PW	2, 7, 28, 30	1000–2000 — 10; 3000–4000 — 13; 5000 — 17	1000 — 1000 PEG; 2000 — 50; — 4000; 3000 — 200; 50 000; 4000 — 200 000; 5000 — 1000; 1 000 000	Polyether with –OH Functionality	Recommended for synthetic water soluble polymers; saccarides, biopolymers of high molecular weight (e.g., lipoproteins), pH range 1–13, spherical, columns of 7.5 mm × 30 and 50 cm
TSK Type SW	2, 7, 28, 30	10	130 (2000 SW) — 2×10^4 PEG; 80 000; 240 (3000 SW) — 4×10^4; 450 (4000 SW) — 2.0×10^5	Silica	pH range 2–8, temperature (max) = 50°C, spherical. The TSK 4000 SW is only available from supplier 51. 2000 SW exclusion limit for protein 100 000; 3000 SW exclusion limit for protein 400 000; 4000 SW exclusion limit for protein 1 000 000 (estimate)
Zorbax SE60–1000	10	6	4×10^4, 7×10^4; 5×10^5, 3×10^6	Silica	Available silanized or unsilanized, compatible with aqueous and non-aqueous solvents
Zorbax PSM 60	10	6 or	2×10^2–10^4	Silica	Available silanized or unsilanized, compatible with aqueous and non-aqueous solvents
Zorbax DSM 1000	10	10	10^4–10^6	Silica	Available silanized or unsilanized, compatible with aqueous and non-aqueous solvents

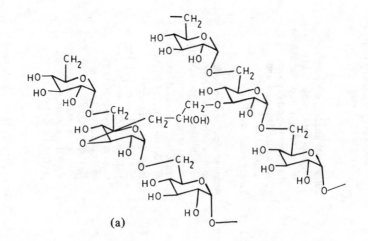

(a)

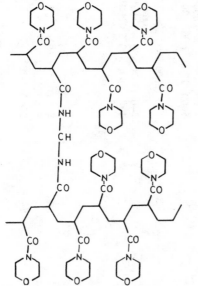

(b)

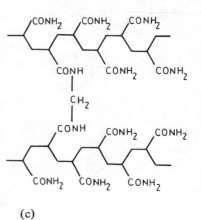

(c)

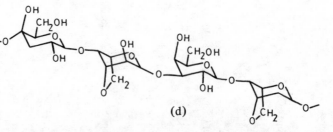

(d)

Fig. 4.8 Exclusion chromatographic supports. (a) Cross-linked dextran (Sephadex). (b) Cross-linked polyacryloylmorpholine (Enzacryl). (c) Cross-linked polyacrylamide (Bio-Gel P). (d) Agarose (linked 1,3 and 1,4 galactose units) (Sepharose).

μm silica also saturates the solvent with silica and does not clog with fines as easily as the 5 μm silica.

In addition to this precolumn it is necessary to have the guard column placed between the injector and the analytical column which will remove any contaminants from dirty samples. It is best not to depend on the guard column to saturate the mobile phase since this length of pellicular bead silica would soon contain voids which would destroy the efficiency of the separation as much as the voids in the microporous particle analytical column.

Drying of solvents can be achieved by the use of three glass columns (2 ft x 1 in (61 x 2.5 cm)) fitted with taps filled with silica gel (12–42 mesh) that has been heated to 180°C for 4 hours. The solvent is passed through three columns in series, discarding the first portion (50 cm^3) to pass through each column, and stored in dark Winchester bottles over Linde molecular sieves. The silica gel in the first column contains any impurities and should be discarded, whilst the silica gel in columns 2 and 3 can be re-activated by drying.

Columns should be stored after washing from them any acids, bases, or salts that might be present.

A number of possible problems in LC with microporous particles is given in Table 4.13[27].

Table 4.13   *Microporous particle column problems* [27].

Problem	Cause	Cure
Increase in inlet pressure	(a) clogged top filter (b) dirty column (c) clogged tubing	(a) replace filter (b) regenerate (c) replace tubing
Loss of resolution or efficiency	(a) dirty column (b) disturbed column bed (c) overloaded sample	(a) regenerate or renew (b) fill gaps with glass beads (c) try smaller samples
Loss of reproducibility	(a) non-equilibrium condition (b) impurities in column (c) wet solvents (d) stabilizers from solvents on column	(a) extend column conditioning period (b) regenerate (c) dry solvents (d) regenerate column and purify solvents
Irregular peak shape	(a) disturbed column bed (b) poor injection (c) sample has decomposed (d) column bed broken	(a) fill gaps with glass beads (b) repeat injection (c) change column packing or use derivatives of sample (d) replace column
Initial specification incorrect (k' and N)	(a) sample injection (b) poor connections (dead volume increased)	(a) syringe injection to column bed gives best efficiency (b) replace tubing

Although pellicular packings are being rapidly replaced by the more efficient microporous particle columns some pellicular columns are still available commercially and there are many useful references to separations involving them in the literature. We feel justified in retaining information on them until such time as their usefulness is exhausted.

References

1. Majors, R. E. (1975) *International Laboratory*, November, p. 13.
1.(a) Majors, R. E. (1980) *J. Chromatog. Sci.*, 18, 488.
2. Horvath, G. C., Preiss, B. A. and Lipsky, S. R. (1967) *Anal. Chem.*, 39, 1422.
3. Williams, R. C., Baker, D. R., Larman, J. P. and Hudson, D. R. (1973) *International Laboratory*, December, p. 39.
4. Endele, R., Halasz, I. and Unger, K. (1974) *J. Chromatog.*, 99, 377.
5. Kirkland, J. J. (1972) *J. Chromatog. Sci.*, 10, 129.
6. Beachell, H. C. and De Stefano, J. J. (1972) *J. Chromatog. Sci.*, 10, 481.
7. Cassidy, R. M., Le Gay, D. S. and Frei, R. W. (1974) *Anal. Chem.*, 46, 340.
8. Bakalyar, S., Yuen, J. and Henry, D. (1975) *Spectra Physics Bulletin.*
9. Snyder, L. R. and Kirkland, J. J. (1974) *Modern Liquid Chromatography*, Wiley, New York.
10. Kirkland, J. J. (1972) *J. Chromatog. Sci.*, 10, 593.
11. Krull, I. S., Wolf, M. H. and Ashworth, R. B. (1978) *International Laboratory*, July, p. 25.
12. Kirkland, J. J. and Antle, P. E. (1977) *J. Chromatog. Sci.*, 15, 137.
13. Leitch, R. E. and De Stefano, J. J. (1973) *J. Chromatog. Sci.*, 11, 105.
14. Huber, J. F. K., Meijers, C. A. M. and Hulsman, J. A. R. J. (1971) *Advances in Chromatography* (ed. A. Zlatkis), Elsevier, New York.
15. Halasz, I. and Sebastian, I. (1969) *Angew. Chem. Internat. Edn.*, 8, 453.
16. Kirkland, J. J. and De Stefano, J. J. (1971) *J. Chromatog. Sci.*, 8, 309.
17. Aue, W. A. and Hastings, C. R. (1969) *J. Chromatog.*, 42, 319.
18. Pryde, A. (1974) *J. Chromatog. Sci.*, 12, 486.
19. Majors, R. E. and Hopper, M. J. (1974) *J. Chromatog. Sci.*, 12, 767.
20. Gray, R. A. C. (1978) *Lab. Equipment Digest*, p. 85.
21. Hastings, C. R., Aue, W. A. and Larsen, F. N. (1971) *J. Chromatog.*, 60, 329.
22. Gilpin, R. K., Korpi, J. A. and Janicki, C. A. (1974) *Anal. Chem.*, 46, 1314.
23. Goldberg, A. (1980) *Liquid Chrom. Tech Reports*, Du Pont.
24. Moore, S., Spackman, D. H. and Stein, W. H. (1958) *Anal. Chem.*, 30, 1185.
25. Knox, J. H. and Vasvari, G. (1974) *J. Chromatog. Sci.*, 12, 449.
26. Epton, R. and Holloway, C. (1973) *Introduction to Permeation Chromatography*, Koch-Light Laboratories.
27. Rabel, F. M. (1975) *International Laboratory*, July, p. 35.

5

Mobile phases in liquid chromatography

5.1 Introduction

In Chapter 4, many different stationary phases were described which are now available commercially. It must be stressed that the suitability of any stationary phase in LC depends on the mobile phase with which it is used; e.g. Durapak columns must *not* be used with aqueous solvents. Some of the qualities that are required in a solvent are now examined.

5.2 Solvent Qualities

The characteristics of a solvent for HPLC include high purity, immiscibility with the stationary phase, absence of reactivity towards the adsorbent, low boiling point, low viscosity, and last, but by no means least, the cost. Some high purity solvents are so expensive that many workers prefer to purchase cheaper solvents to be purified in their own laboratory.

5.2.1 Purification

The purification or clean-up procedure must include a filtration step using a millipore system if need be. Distillation of the solvent from glass removes many of the impurities, e.g. those with ultraviolet chromophores, whilst passage over alumina or silica may remove more polar constituents. Removal of bacteria and organic compounds are the two major difficulties in purifying water. Some solvents are sold with a preservative; e.g. chloroform contains methanol which must be removed before the solvent is pure enough for LC. In such cases, passage of the solvent through alumina or silica activated at 125°C may be necessary. Alternatively, preservative-free solvents have their own chemical hazards, so care must be taken in their use; e.g. chloroform may produce phosgene which is highly toxic, and very pure diethyl ether, if not stored correctly, may produce peroxides which are explosion risks.

5.2.2 Stability

When acetone is used with basic alumina, diacetone alcohol is produced by a base-catalysed self-condensation reaction. It is clear that these two components, acetone and basic alumina, are not compatible, because the eluent from the column will be contaminated with diacetone alcohol and produce, at best, extraneous base-line drift and possibly a second peak. The solvent must not be decomposed by the stationary phase, and equally the stationary phase should not be attacked by the mobile one. In Chapter 4 it was explained that one of the driving forces for chemically bound mobile phases was the difficulty which had existed previously with high loadings of stationary phases. Using a mechanically held stationary phase, the mobile phase should be completely immiscible with the stationary phase, but in practice most mobile phases dissolve the stationary phase to some extent. The solvent then has to be pre-saturated with the stationary phase before reaching the column. The pre-column, the solvent, and the stationary phase must all be maintained at the same temperature. Such a technique is awkward and exemplifies one of the properties required of the solvent, i.e. that it should *not* change during the analysis.

The water content of alumina or silica has to be held within prescribed limits to control its activity. If the mobile phase is dry and it removes the water, an unacceptable variation is introduced. It is sometimes necessary to pre-saturate a dry solvent with water so that reproducible k' values can be obtained. A simple method of pre-saturation is to pass the dry solvent through a glass column packed with silica gel containing a high loading of water (20–30%), if need be several times to reach saturation level. The optimum water content is usually 25% water saturated solvent for alumina and 50% for silica [1].

Solvents often need to be 'degassed'. The presence of dissolved oxygen can lead to the formation of bubbles in the detector or the connecting tubing. In addition, dissolved oxygen can cause a significant change in the level of absorbance at 210 nm with methanol. Of the methods for removing these gaseous impurities, ultrasonic degassing has been shown to be ineffective and can lead to increased oxygen levels. Only refluxing is completely effective in removing oxygen from all solvents, but the solvent has to be kept warm to prevent subsequent re-adsorption. Helium replacement (or sparging) and vacuum degassing are only moderately effective in removing oxygen from solvents such as water but they are very good for hexane. With methanol and acetonitrile these two methods are good enough to reduce bubble formation at the inlet to a pump but do not provide the lowest u.v. absorption.

When oxygen is removed it is necessary to prevent re-absorption since hexane can take up 90% of its original oxygen content within 20 min at room temperature.

5.2.3   Detector Problems

The ultraviolet detector can only monitor the presence of oligomers of polystyrene at 254 nm because the aromatic ring has a strong u.v. chromophore at 254 nm. The use of benzene for the elution would be unacceptable because it too absorbs at 254 nm. The u.v. cut-off for many common solvents is given in Table 5.1. Gradient elution with u.v. detectors can also be troublesome if either of the mixed solvents has any absorptivity.

Table 5.1   *Properties of solvents (Eluotropic series)*

	$\epsilon^{\circ}_{Al_2O_3}$	RI	U.v. cut-off (nm)	Viscosity (cP) at 20°	BP (°C)	δ	P'
n-Pentane	0.00	1.358	210	0.23	36.0	7.1	0.0
n-Hexane	0.01	1.375	210		68.7	7.3	0.0
n-Heptane	0.01	1.388	210	0.41	98.4	7.4	0.0
Cyclohexane	0.04	1.427	210	1.00	81.0	8.2	0.0
Carbon disulphide	0.15	1.626	380	0.37	45.0	10.0	1.0
Carbon tetrachloride	0.18	1.466	265	0.97	76.7	8.6	1.7
Isopropyl ether	0.28	1.368	220	0.37	69.0	7.0	2.2
2-Chloropropane	0.29	1.378	225	0.33	34.8		
Toluene	0.29	1.496	285	0.59	110.6	8.9	2.3
1-Chloropropane	0.30	1.389	225	0.35	46.6	8.3	
Chlorobenzene	0.30	1.525	280	0.80	132.0	9.6	2.7
Benzene	0.32	1.501	280	0.65	80.1	9.2	3.0
Bromoethane	0.37	1.424	225	0.41	38.4	8.8	3.1
Ethyl ether	0.38	1.353	220	0.23	34.6	7.4	2.9
Chloroform	0.40	1.443	245	0.57	61.2	9.1	4.4
Dichloromethane	0.42	1.424	245	0.44	41.0	9.6	3.4
Tetrahydrofuran	0.45	1.408	222	0.55	65.0	9.1	4.2
1,2-Dichloroethane	0.49	1.445	230	0.79	84.0	9.7	3.7
Methyl ethyl ketone	0.51	1.381	330	0.43	79.6		4.5
Acetone	0.56	1.359	330	0.32	56.2	9.4	5.4
Dioxan	0.56	1.422	220	1.54	104.0	9.8	4.8
Ethyl acetate	0.58	1.370	260	0.45	77.1		4.3
Methyl acetate	0.60	1.362	260	0.37	57.0	9.2	
Pentan-1-ol	0.61	1.410	210	4.10	137.3		
Dimethyl sulphoxide	0.62	1.478	270	2.24	190.0	12.8	6.5
Aniline	0.62	1.586	325	4.40	184.0		6.2
Nitromethane	0.64	1.394	380	0.67	100.8	11.0	6.8
Acetonitrile	0.65	1.344	210	0.37	80.1	11.8	6.2
Pyridine	0.71	1.510	305	0.94	115.5	10.4	5.3
Propan-2-ol	0.82	1.380	210	2.30	82.4	10.2	4.3
Ethanol	0.88	1.361	210	1.20	78.5	11.2	5.2
Methanol	0.95	1.329	210	0.60	65.0	12.9	6.6
Ethylene glycol	1.11	1.427	210	19.90	198.0	14.7	5.4
Acetic acid		1.372	251	1.26	118.5	12.4	6.2

$\epsilon_{Al_2O_3}$	Snyder's eluent strength function
RI	Refractive index, n_D
U.v. cut-off	Wavelength below which the solvent cannot be used
Viscosity	in centipoises
BP	Boiling point in °C
δ	Hildebrand solubility parameter
P'	Polarity index calculated from Rohrschneider's data

The moving wire or chain detectors are not troubled by pure solvents but any particulate matter on the chain will produce noise spikes. Thus solvents with inorganic ions can cause difficulties.

Refractometers depend on the difference in refractive index between the solute and the mobile phase. If the solvent and solute have only slightly different refractive indices, the ability to detect the solute will be reduced.

5.2.4 Solubility and Viscosity Effects

The solvent must be capable of dissolving the solutes. This fact may lead to difficulties if the solutes contain differing numbers of functional groups; e.g. lipid mixtures often contain hydrocarbons, wax esters, triglycerides, and monoglycerides which will dissolve only in a polar solvent. Such a solvent used as the mobile phase will remove the hydrocarbons and wax esters together in the first column volume after V_0 [2]. In this situation it may be necessary to accept the poor separation and re-chromatograph these two lipid classes in a less polar solvent in which they will dissolve but which will not dissolve the monoglycerides. This is an example of the general elution problem which will be considered in Chapter 6.

Given a choice between two solvents of similar polarity, the less viscous is always chosen, k' is proportional to solvent viscosity in LSC. A viscous solvent cuts down column permeability and higher pressures need to be used. In addition, diffusion and mass transfer effects are reduced. Solvents with viscosities in the range 0.4–0.5 centipoise are usually best. Indeed, very low viscosities, e.g. 0.2 centipoise, are associated with low boiling points, and solvents with these properties tend to form bubbles in the column. A doubling of solvent viscosity results in a doubling of separation time if the efficiency remains the same.

5.3 Liquid–Solid Chromatography

5.3.1 Eluotropic Series

Early in the history of column chromatography, the eluting power of a solvent, i.e. its effectiveness in pushing solutes from an adsorbent, was recognized. The ordered presentation of solvents is called an eluotropic series. Trappe [3], in 1940, established an eluotropic series of solvents useful for lipids, and in 1942, Strain [4] published an expanded list with essentially the same order. Jacques and Mathieu [5] claimed that the eluting power of a solvent was proportional to its dielectric constant, and for many years this empirical relationship was used to place solvents in an eluotropic series. More recently, the order of an eluotropic series has been found to correspond better to Snyder's 'eluent strength function' which is defined as the adsorption energy per unit area of standard activity of the solvent [6].

$$\log K^\circ = \log V_a + E_a(S^\circ - A_S \epsilon^\circ) \tag{5.1}$$

where V_a is the adsorbent surface (volume of solvent monolayer approximately 0.00035 x surface area); E_a is the adsorbent energy function which is proportional to the average surface energy of adsorbent; S^o is the adsorption energy of the solute; A_S is the area of solid required by the adsorbed solute.

This equation shows how the distribution coefficient K is related to the adsorbent surface area and the surface energy. If Equation (5.1) is applied to two different analyses, of the same solute on the same adsorbent with two mobile phases, it is found that:

$$\log [(K^o)_1/(K^o)_2] = E_a A_S(\epsilon^o_2 - \epsilon^o_1) \qquad (5.2)$$

i.e. the elution volumes of the solute would depend on the differences in eluent strength of the two mobile phases. The eluent strength of a solvent is an attempt to put a numerical value on the term 'polarity' so beloved by chemists. It is clear that the greater the eluent strength the smaller is K^o for a particular solute and adsorbent. ϵ^o is arbitrarily defined as zero on alumina when n-pentane is used as a solvent. Snyder claims that an ϵ^o value on silica is equal to 0.77 of the value on alumina.

Thus it can be seen (Table 5.1) that hydrocarbons have low values of ϵ^o (n-heptane 0.01), ethers have medium values (isopropyl ether 0.28), and alcohols have very high values (methanol 0.95).

The advantage of an eluotropic series is that it enables the choice of solvent to be slightly less empirical than previously. Preliminary experiments are usually performed on thin-layer chromatography with one solvent, say cyclohexane (ϵ^o 0.04). If cyclohexane proves to be too weak, the solute will bind to the adsorbent too well (k' will be too large). In the competitive equilibrium between solute and solvent molecules for sites on the adsorbent surface, the solvent molecules are first adsorbed on to the adsorbent surface:

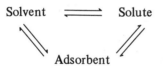

$$\text{Solute}_{mobile} + nC_6H_{12\ adsorbed} = \text{Solute}_{adsorbed} + nC_6H_{12\ mobile}$$

The solute molecules in the mobile phase must first displace the cyclohexane molecules from the adsorbent surface. The net energy of adsorption ΔE_a determines the ratio of adsorbed to non-adsorbed solute molecules. ΔE_a is approximately equal to the energy of the adsorbed solute molecules less n times the energy of the adsorbed solvent molecules. The interaction energies are determined by the interaction of solute and solvent with the adsorbent surface, so the energy of the adsorbed solute molecules is not a function of the solvent, nor is the energy of the adsorbed solvent dependent on the solute. The effect of the solvent on ΔE_a, i.e. the solvent strength, is a function only of the interaction energy of the adsorbed solvent molecules and is independent of the solute molecules. ϵ° is a function only of the solvent and the adsorbent.

When cyclohexane is too weak (k' too large) the equilibrium lies very far to the right-hand side.

A second trial TLC, with dichloromethane (ϵ° 0.42) as solvent, would result in a new equilibrium:

$$\text{Solute}_{\text{mobile}} + n\text{CH}_2\text{Cl}_2\text{ adsorbed} = \text{Solute}_{\text{adsorbed}} + n\text{CH}_2\text{Cl}_2\text{ mobile}$$

which might be too far to the left-hand side. This results in too small a k' value, i.e. the solute would pass through the column too quickly. For the next trial, a pure solvent with ϵ° between 0.04 and 0.42 would be substituted, or a mixture of dichloromethane and cyclohexane to give a suitable ϵ° value.

Temperature is not a variable that creates many problems in LSC since k' values decrease 2% per degree with increasing temperature.

5.3.2 Binary Mixtures

Several binary mixtures are given in Tables 5.2 and 5.3. Whilst most eluotropic series relate the solvent to alumina, a similar order prevails for silica but the

Table 5.2   *Eluent strengths of binary mixtures on alumina* [1].

2-Chloropropane–Pentane		Eluent strength, ϵ	Diethyl ether–Pentane	
8	92	0.05	4	96
19	81	0.10	9	91
34	66	0.15	15	85
52	48	0.20	25	75
77	23	0.25	38	62
Dichloromethane–Pentane			Diethyl ether–Pentane	
13	87	0.20	25	75
22	78	0.25	38	62
34	66	0.30	55	45
54	46	0.35	81	19
Methanol–Diethyl ether			Methyl acetate–Pentane	
2	98	0.40	19	81
4	96	0.45	29	71
8	92	0.50	44	56
20	80	0.55	65	35
50	50	0.60		

Table 5.3   *Eluent strengths of binary mixtures on silica.*

Methanol–Diethyl ether		Eluent strength, ϵ	Acetonitrile–Methanol	
0.25	99.75	0.40		
0.75	99.25	0.45		
1.70	98.30	0.50		
3.50	96.50	0.55		
8.00	92.00	0.60		
18.00	82.00	0.65	70	30
42.00	58.00	0.70	60	40
100.00	0.00	0.73	0	100

order is reversed for graphitized carbon. It is possible to obtain a gradual change in eluent strength by choosing suitable binary mixtures of solvents; e.g. pentane ($\epsilon°$ 0.00) and 2-chloropropane ($\epsilon°$ 0.29) give a range of eluent strengths from 0.00 at 100% pentane to 0.29 at 100% 2-chloropropane.

The figures in Tables 5.2 and 5.3 explain the simple rule-of-thumb that has been used for many years in the situation where the solute mixture is completely unknown; i.e. the chromatogram is started with a low polarity solvent and then a more polar solvent is added in increasing steps 2, 4, 8, 16, 32%. It can be seen from the methanol–ether mixtures that each step corresponds approximately to an increase in eluent strength of 0.05 unit.

One advantage of using binary mixtures of solvents is that increased eluent strength can be obtained without increasing the viscosity. Ethylene glycol is very viscous (19.9 cP) and outside the range recommended earlier. Its use as a mobile phase would give greatly enhanced eluent strength but its viscosity would necessitate the use of excessively high pressures. If ethylene glycol were used at the 1 or 2% level in a much less polar solvent, a compromise could be reached between eluent strength and pumping pressure.

The other advantage of binary solvents relates to the problem where two bands are incompletely resolved though the k' is the best for any two adjacent bands. The resolution, which is measured by the distance between the two band centres divided by the average band width, is related to the separation selectivity, the column efficiency, and the band migration rates:

$$R_S = \frac{1}{4}\left(\frac{\alpha - 1}{\alpha}\right)\left(\frac{k'}{1 + k'}\right)(N)^{\frac{1}{2}} \qquad (2.58)$$

The resolution can be controlled by varying the separation factor (α), the column plate number (N), or the capacity factor, each of which may be considered independently and each of which is affected by the solvent. When two bands are not resolved though k' has been optimized, an improvement can be achieved either by increasing column efficiency or by changing the separation selectivity. A trial-and-error method can be used when two solutes have an α value approximately 1.0 for a binary mixture, 55% ether in pentane. The eluent strength of 55% ether in pentane is 0.30. Another solvent mixture (8% methyl acetate in pentane) has the same eluent strength (0.30) but its different composition should change the separation selectivity. Instead of this empirical approach, we can try to vary α by using some general rules.

The maximum value of α is obtained when we have either a large concentration of the more polar solvent or a small concentration of the more polar solvent [7]. To obtain the greatest selectivity, the solvent should be capable of hydrogen bonding to the solute; for example, propan-2-ol in benzene would be better than acetonitrile in benzene.

One of the disadvantages of solvent mixtures is that the ϵ° values are not so predictable as the equivalent polarity index which is used in LLC. A second disadvantage of mixtures is that there is a possibility of solvent de-mixing. This problem, often noted in TLC where ternary mixtures are used, leads to two solvent fronts. In HPLC even the binary mixture propanol in hexane shows the phenomenon where the strong solvent, propanol, is preferentially adsorbed at the head of the column, whilst pure hexane, the weak solvent, continues to flow through the column. This problem is overcome by passing enough of the solvent through the column for the whole column to reach equilibrium.

A further disadvantage of binary mixtures concerns the use of solvents of high eluent strength. If the column is to be used under new isocratic conditions subsequently, it may be necessary to regenerate the column by passage of solvents of gradually decreasing eluent strength to remove the solvent of high ϵ° which would otherwise stay on the column.

5.3.3   Reconditioning

Finally there is the problem of reconditioning a column packing that has been used for a long time with dirty samples or solvents. Impurities of high polarity will remain adsorbed on the head of the column packing and can ultimately cause non-reproducible results. In order to remove the polar compounds it is often necessary to pass methanol (ϵ° 0.95) through the column. Because methanol has such a large solvent strength parameter, it is then necessary to wash out all the adsorbed methanol by passing ten column volumes of a less polar solvent with which methanol is miscible, as for binary mixtures (Section 5.3.2). Solvents that have been suitable for column reconditioning are methanol, acetone, ethyl acetate, trichloroethane, and heptane.

5.4   Liquid–Liquid Chromatography

In liquid–liquid chromatography, it is not surprising that many of the criteria desired by Hais and Macek [8] in 1963 for paper chromatography apply equally for HPLC (1) k' values should be between 2 and 5. (2) In the chosen system of solute, solvent, and stationary phase, the partition isotherm should be linear and peaks should be symmetrical, unaffected by solvent concentration. (3) The system should *not* cause any chemical changes in the solutes. (4) The solvents should *not* react with the detection agent or reduce detector sensitivity, but any enhancement of sensitivity would be an advantage.

One factor that is not appropriate for paper chromatography is the immiscibility of the stationary phase and the mobile phase. This feature has been touched on in Chapter 4, where it was accepted that it was not possible to get two phases that are completely immiscible. Pre-saturation of the mobile phase can be accomplished by equilibrating the solvent with the stationary phase in a separating funnel. However, a better technique is to place a large excess of the mobile phase in a large conical flask with the stationary phase, and blow nitrogen through the mixture to remove any dissolved oxygen. The conical flask is then placed on a magnetic stirrer and the mixture kept agitated overnight. The mobile phase, now saturated with the stationary phase, is separated off and stored under nitrogen until used. In order to limit the chance of the solvent removing the stationary phase, a pre-column is often used before the main analytical column. This pre-column contains a coarse, wide mesh support which has been coated with 20–30% of the stationary phase. It should be renewed frequently to ensure that the solvent is saturated with the stationary phase at all times.

Separations on LLC depend on the nature of both the mobile phase (solvent) and the stationary phase but as usual with LC it is best to vary k' by changing the solvent. Eluotropic series can again be set up (Table 5.4) using the

Table 5.4   *Properties of solvents (solubility parameters and polarity index)*

Solvent	δ	London forces	Polar forces	Hydrogen bonding	P'	x_e	x_d	x_n	Dipole moment
Hexane	7.2	7.2	0.0	0.0	0.0	0.0	0.0	0.0	0.00
Diethyl ether	7.6	7.1	1.4	2.5	2.9	0.6	0.1	0.3	0.34
Carbon tetrachloride	8.7	8.7	0.0	0.0	1.7	0.3	0.4	0.3	0.32
Toluene	8.9	8.8	0.7	1.0	2.3	0.3	0.2	0.4	0.44
Ethyl acetate	9.1	7.4	2.6	4.5	4.3	0.3	0.3	0.4	0.42
Benzene	9.2	9.0	0.5	1.0	3.0	0.3	0.3	0.4	0.43
Chloroform	9.2	8.7	1.5	2.8	4.4	0.3	0.4	0.3	0.40
Acetone	9.8	7.6	5.1	3.4	5.4	0.4	0.2	0.4	0.40
Dichloromethane	9.9	8.9	3.1	3.0	3.4	0.3	0.2	0.5	0.49
Acetic acid	10.5	7.1	3.9	6.6	6.2	0.4	0.3	0.3	
Butan-1-ol	11.3	7.8	2.8	7.7	3.9	0.5	0.2	0.3	0.26
Acetonitrile	11.9	7.5	8.8	3.0	6.2	0.3	0.3	0.4	0.41
Ethanol	12.9	7.7	4.3	9.5	5.2	0.5	0.2	0.3	0.28
Methanol	14.3	7.4	6.0	10.9	6.6	0.5	0.2	0.3	0.30
Ethylene glycol	16.3	8.3	5.4	12.7	5.4	0.5	0.2	0.3	0.30
Water	23.3	6.0	15.3	16.7	9.0	0.4	0.3	0.3	0.26

δ Solubility parameter
P' Polarity index
x_e Selectivity parameter (proton acceptor)
x_d Selectivity parameter (proton donor)
x_n Selectivity parameter (interaction with dipole)

Hildebrand solubility parameter δ [9] to give a more quantitative measure. This solubility parameter is a good approximation to what is called polarity; i.e. very polar solvents such as methanol have high values (12.9) whilst non-polar solvents such as fluoroalkanes have low values (6.0).

The solubility parameter is used by most workers merely to indicate the relative position of a solvent in an eluotropic series. It seems likely, however, that substantial progress towards a less empirical approach to solvent selection will only occur when the use of the solubility parameters has become widespread.

The solubility parameter δ can be shown to be made up of four parts, concerned with specific intermolecular interactions, e.g. dispersion, dipole orientation, and hydrogen bonding [10]:

$$\delta = \delta_d + \delta_0 + \delta_a + \delta_h \tag{5.3}$$

where δ_d is a measure of how the solvent interacts with solute molecules by means of London dispersion forces; δ_0 is a measure of how the solvent interacts by dipole interaction; δ_a is a measure of the ability of the solvent to interact as a hydrogen acceptor; δ_h is a measure of its interaction as a hydrogen donor.

The solvent strength is governed by the Hildebrand solubility parameter but the selectivity is controlled by the dispersion forces, the dipole interactions, and the hydrogen donor and acceptor parts of δ. A short list of solvents and their Hildebrand solubility parameters is given in Table 5.4; a fuller list is provided by Snyder and Kirkland [1] and by Waters Associates [11].

One disadvantage of the eluotropic series based on δ is highlighted by the position of diethyl ether which appears at the same place as hexane although it is a much more polar solvent than hexane. Snyder [12] has made use of Rohrschneider's solubility data [13] to produce a solvent characterization parameter P' which does not depend on the 'present less-than-ideal theory of liquid mixtures':

$$P' = \log(K_g'')_{\text{ethanol}} + \log(K_g'')_{\text{dioxan}} + \log(K_g'')_{\text{nitromethane}} \tag{5.4}$$

Log K_g'' is proportional to the free energy of vaporization of the solvent, and is calculated from solubility data for ethanol, dioxan, and nitromethane for each solvent. This scheme distinguishes between solvent strength, which is its ability to dissolve more polar molecules preferentially, and solvent selectivity, which is its ability to dissolve one compound selectively as opposed to another where the polarities are similar. Snyder defined selectivity parameters:

$$x_e = \log(K_g'')_{\text{ethanol}}/P'$$
$$x_d = \log(K_g'')_{\text{dioxan}}/P'$$
$$x_n = \log(K_g'')_{\text{nitromethane}}/P'$$

and he suggested that they could be considered to reflect the ability of the solvent to act as a proton acceptor (x_e), or as a proton donor (x_d), or by a strong dipole interaction (x_n). Diethyl ether is a good proton acceptor, and a polar solvent compared with hexane. It shows a low δ value because δ values are measured for the pure liquid where no proton donor molecules are present. On the P' scale it has a much higher value (2.9) than hydrocarbon solvents (0.00), which is a better measure of its polarity.

The polarity index is the variable to alter when adjusting k', but the selectivity (α) is changed by altering the solvent to another group. Snyder [12] classifies solvents into groups on the basis that similar classes of solvent will behave similarly (Table 5.5). Solvent selection using Tables 5.4 and 5.5 is dependent on choosing a solvent in which the sample is soluble, e.g. chloroform. If the k' value of the solute proves to be too low, it is noted that hexane has low values of P' (Table 5.4). Using a chloroform—hexane (10 : 90) mixture, the k' value is in the correct range between 1 and 10 but the resolution is not satisfactory ($\alpha \approx 1.0$). The polarity index for this chloroform—hexane mixture is approximately 0.4, so we must choose another solvent, from a different group, that will maintain the polarity index at about 0.4. Chloroform belongs to group 9, so a solvent from a group that is as far away as possible is chosen so as to be chemically different, e.g. group 1. Isopropyl ether at a 20% level in hexane will give a polarity index of 0.4 and thus the same k' values. If resolution has not been improved, 12% dichloromethane in hexane will give similar k' values and could be used.

Table 5.5   *Group classification of solvents.*

Group	Solvents
1	Aliphatic ethers, trialkylamines, tetramethylguanidine
2	Aliphatic alcohols
3	Pyridines, tetrahydrofuran, amides (except formamide)
4	Glycols, glycol ethers, benzyl alcohol, formamide, acetic acid
5	Dichloromethane, dichloroethylene, tricresyl phosphate
6	Halogenated alkanes, ketones, esters, nitriles, sulphoxides, sulphones, aniline, dioxan
7	Nitro-compounds, propylene carbonate, phenylalkyl ethers, aromatic hydrocarbons
8	Halogenobenzenes, diphenyl ether
9	Fluoroalkanols, *m*-cresol, chloroform, water

5.4.1   Solvents for Reversed-Phase Chromatography

Ion Suppression. Where the packing has an octadecyl or an octyl chain, the solvent is normally a polar one, i.e. this is the reversal of polar solid support and non-polar solvent. Water mixed with methanol, acetonitrile, dioxan or tetra-hydrofuran is the most popular polar solvent. Polar solutes are eluted first because they spend more time in the mobile phase. Some polar solutes especially ions can give poor shaped peaks or can be eluted too quickly. If a short-chain carboxylic acid, e.g. butanoic, is analysed, it may exist as its carboxylate anion and give a badly shaped peak.

$$CH_3(CH_2)_2CO_2H \rightleftharpoons CH_3(CH_2)_2CO_2^- + H^+$$

By making the solvent more acidic, the ionization can be suppressed and the butanoic acid separates by utilizing the reverse-phase conditions to the full. This technique of modifying the solvent is known as 'ion suppression'.

Reverse Phase Ion Partition. Ion suppression cannot be accomplished in the pH range 2–8 because strong acids are still completely ionized. If a counter-ion is added to the solvent, it is possible to convert all the solute ions into ion-pairs which, being electrically neutral, can also be analysed in the reverse phase mode. Thus an amine can be paired with a sulphonate:

$$RNH_3^+ Cl^- + CH_3(CH_2)_6 SO_3^- Na^+ \rightleftharpoons RNH_3^+ \overline{SO}_3(CH_2)_6 CH_3 + Na^+ + Cl^-$$

or a carboxylic acid as its quarternary ammonium salt e.g. cetrimide:

$$RCO_2^- \overset{+}{Na} + R_4'N^+ \overline{Cl} \rightleftharpoons RCO_2^- \overset{+}{N} R_4' + Na^+ + Cl^-$$

The effect of reducing the chain length in the sulphonic acid [14] was to reduce the retention e.g. LSD and iso-LSD which eluted in 24 and 32 min with the C_7 sulphonic acid were separated at 5 and 8 min when methane sulphonate was used as the counter-ion.

5.4.2 Solvents for Normal-Phase Chromatography

Where the bonded phase has been prepared with polar functional groups, e.g. NH_2, nitrile, fluoroether or diol (Table 4.5) the use of less hydroxylic solvents can result in normal-phase chromatography. This technique may give results similar to the separations on silica or alumina, but the absence of silanols on the adsorbent reduces tailing. So normal phase chromatography can be applied to silica packings or to these polar phases.

5.4.3 Properties of Specific Solvents

Acetonitrile is one of the most popular reverse phase solvents because it has a low viscosity, favourable vapour pressure and u.v. transparency. One of its disadvantages is the 40 p.p.m. threshold limit value which makes it one of the most toxic HPLC solvents. It can be purified by drying over P_2O_5. Of the alcohols, methanol is the most widely used because the higher homologues are very viscous; higher homologues can sometimes be useful however, since the latter are miscible with both polar and non-polar solvents. Ethanol can be dried by distillation from magnesium turnings.

Tetrahydrofuran is less polar than methanol or acetonitrile with the result that less tetrahydrofuran needs to be used to produce equivalent retention in reverse phase. Tetrahydrofuran and the other ethers can form peroxides which may cause explosions. Diethyl ether has such a low vapour pressure and a low flash point that it is better not to use it in HPLC. Tetrahydrofuran is purified by treatment with potassium hydroxide, then lithium aluminium hydride and finally distilled.

Of the hydrocarbon solvents, pentane, hexane, heptane and iso-octane have almost identical chemical properties. Pentane may be used for its low viscosity but it has a high vapour pressure and a very high compressibility. Hexane, heptane and iso-octane can be be used interchangeably with the choice depending on price. Heptane can be obtained 99% pure whereas hexane is purchased as a mixture of isomers with n– C_6 present at 95%. Cyclohexane has the highest viscosity and the unusual property of freezing at 5000 lbf in^{-2} (340 atm) with consequent damage to the pumps. Saturated hydrocarbons are purified by washing with conc. sulphuric acid and then distilled.

Dichloromethane and chloroform are the most important of the chlorinated solvents which all decompose in air and moisture to give hydrogen chloride. It must be remembered that chloroform is often sold with one to two per cent ethanol as a stabilizer and any small change in this concentration can alter the solvent strength considerably. Dichloromethane is less toxic than chloroform and has a u.v. cut-off at 230 nm against 245 nm for chloroform, but it has a higher vapour pressure which may lead to cavitation in the pump. Carbon tetrachloride is more toxic than either dichloromethane or chloroform with no special advantages. Chloroform and dichloromethane are treated with conc. sulphuric acid then water and dried over calcium chloride to purify them.

Most of the other solvents, e.g. acetone, methyl ethyl ketone, ethyl acetate, dimethylsulphoxide, dimethylformamide and carbon disulphide have u.v. cut-off values above 254 nm and are used only when their special solubility is required.

5.5 Ion-Exchange Chromatography

The solvent most widely used in ion exchange chromatography is water. Indeed it is only recently that ion exchangers that can withstand organic solvents have become available. Changes of k' in ion-exchange chromatography can be effected by altering the pH of the solvent or by changing the ionic strength of the mobile phase.

As pH is increased, so the retention volume of a solute decreases in cation exchange chromatography. Thus, at higher pH values, a carboxylic acid in equilibrium with its anion and hydrogen ion will have a higher concentration of carboxylic ions:

$$-COOH \rightleftharpoons -COO^- + H^+$$

The solute ion is retained more strongly because it can compete more strongly for anion sites in the resin at higher pH values. Conversely, at lower pH values the cation of a weak base is formed more readily:

$$-NH + H^+ \rightleftharpoons -\overset{+}{N}H_2$$

Thus, lowering the pH increases the retention volume of a solute because there is increased competition of the charged solute ions for the cation exchange sites.

Buffer solutions are used to control the pH in ion-exchange chromatography, i.e. ammonia and pyridine and other cationic buffers for anion exchangers, or acetate, formate, citrate, and other anionic buffers for cation exchange resins. The retention of solute molecules decreases as the concentration of the salt in the mobile phase increases, since the sample ions cannot compete so well with the solvent counter-ions for the ion exchange groups in the resin.

The effectiveness of several ions to push solute ions from their respective exchangers is given in Table 5.6. The order shows that citrate is very strongly bonded to the resin, which means that the citrate ions elute sample ions more quickly than fluoride.

Table 5.6   *Retention sequence for ions in ion exchange chromatography.*

Anion exchangers	Cation exchangers
Citrate	Ba^{2+}
Sulphate	Pb^{2+}
Oxalate	Sr^{2+}
Iodide	Ca^{2+}
Nitrate	Ni^{2+}
Chromate	Cd^{2+}
Bromide	Cu^{2+}
Thiocyanide	Co^{2+}
Chloride	Zn^{2+}
Formate	Mg^{2+}
Acetate	UO_2^{2+}
Hydroxide	Te^{+}
Fluoride	Ag^{+}
	Cs^{+}
	Rb^{+}
	K^{+}
	NH_4^{+}
	Na^{+}
	H^{+}
	Li^{+}

5.6   Exclusion Chromatography

The mobile phases in exclusion chromatography are chosen for their ability to dissolve the solute molecules and not for their control of resolution. The solvent should not attack the column packing but should have a low viscosity at the operating temperature. All exclusion chromatography packings will shrink and swell depending on the solvents used. Too much shrinking will lead to spaces which will destroy the efficiency of the column. Some commonly used exclusion chromatography solvents are given in Table 5.7.

Table 5.7 *Properties of solvents for exclusion chromatography.*

Solvent	U.v. cut-off (nm)	Viscosity (cP) at $20°C$	Max. flow $cm^3 min^{-1}$	Refractive index
Hexane	210	0.326	7.2	1.3749
Tetrahydrofuran	220	0.550	4.0	1.4072
Dichloromethane	220	0.440	5.3	1.4237
p-Dioxan	220	1.439	1.6	1.4221
Cyclohexane	220	0.980	2.4	1.4262
Dichloroethane	225	0.840	2.8	1.4444
Trichloroethane	225	1.200	2.0	1.4791
Chloroform	245	0.580	4.0	1.4457
Carbon tetrachloride	265	0.969	2.4	1.4630
Benzene	280	0.652	3.6	1.5011
Toluene	285	0.590	4.0	1.4969
Xylene	290	0.810	2.9	1.4972
N,N,-Dimethylformamide	295	0.924	2.5	1.4294
m-Cresol*		184.200		1.5400
1,2,4-Trichlorobenzene*		0.50	4.0	1.5700

* Standard high temperature solvents

5.7 Gradient Elution

A major contribution to the range of separations that can be successfully achieved by liquid chromatography is that of solvent programming or gradient elution which is performed by changing the composition of the mobile phase by programming a strong solvent with increasing concentration into a weak initial solvent. As the solvent strength changes, so the k' values of the solute bands alter, allowing all the bands to be eluted in the optimum k' range $1 < k' < 10$. The simplest way of varying the solvent composition is the one that has been used since the early days of liquid chromatography, namely stepwise elution. The solvent in the reservoir is changed during the course of the analysis by introducing fresh batches of solvent of differing eluent strength. The major disadvantages tend to be that the overall analysis time is increased and that the sharp changes in eluent strength can lead to additional problems, e.g. with the detector.

A more suitable way to vary the solvent composition is to use gradient elution with two solvents – a weak solvent 1 and a stronger solvent 2. The chromatogram is started with pure solvent 1 which must be weak enough to give good resolution of the solutes with low k' values eluting at the start of the chromatogram. The addition of solvent 2 to solvent 1 varies the eluent strength until a concentration of pure solvent 2 is reached. Solvent 2 must be strong enough to elute the strongly retained solutes with large k' values in a reasonable time. The solvent programme can be linear, convex, or concave as shown in Fig. 5.1.

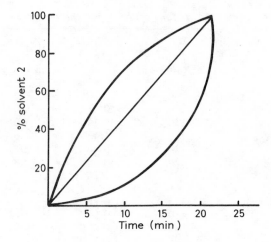

Fig. 5.1 Solvent programmes for gradient elution.

For optimum results in LSC, the concentration of solvent 2 in the eluent should increase exponentially with time, i.e. concave in Fig. 5.1. For LLC with bonded phases, it seems likely that a linear programme should be adopted.

Fig. 5.2 shows some of the results that can be achieved with gradient elution. One of the commonest faults in this technique is to start with a solvent that has too high an eluent strength. The early bands are therefore pushed from the column all together [Fig. 5.2(a)].

If the second solvent has too low an eluent strength, or if the final concentration of solvent 2 in 1 is too low, the solutes continue to be eluted after the final concentration has been reached, i.e. isocratic conditions. In this case, the band widths increase and some components may not be eluted from the column at all [Fig. 5.2(b)].

If the eluent strength of solvent 1 is very different from that of solvent 2, i.e. it is greater than is required, all the bands become crowded together in the middle of the chromatogram and the resolution is poorer [Fig. 5.2(c)].

When solvents 1 and 2 have eluent strengths that are too similar, the gradient elution separation appears like an isocratic elution, i.e. the early bands are poorly resolved whilst later bands are too wide and have long retention times [Fig. 5.2(d)].

The phenomenon of solvent de-mixing is exemplified in Fig. 5.2(e). The solvent strength of solvent 2 is too different from that of solvent 1, with the result that solvent 2 is held at the inlet end of the column at the beginning of the gradient elution and pure solvent 1 flows through the column. The early peaks (a, b, c) show isocratic elution, then solvent 2 breaks through the column when it has become saturated, and peaks d, e, f are displaced with poor resolution. Finally peaks g, h, i are eluted with gradient elution.

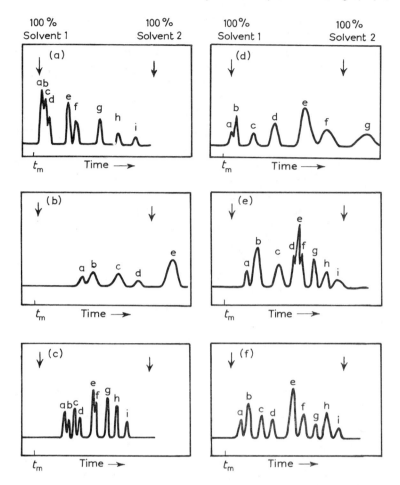

Fig. 5.2 Gradient elution chromatograms. (a) $\epsilon°$ of Solvent 1 is too high. (b) $\epsilon°$ of Solvent 2 is too low. (c) $\epsilon°$ of Solvents 1 and 2 are very different. (d) $\epsilon°$ of Solvents 1 and 2 are too similar. (e) Solvent demixing. (f) Optimum solvent programme.

When all of these pitfalls have been avoided, when solvents 1 and 2 are the most suitable solvents, and the programme is optimized, the chromatogram should look like Fig. 5.2(f).

Gradient elution has a major disadvantage in that it cannot be used with the refractive index detector because it is not easy to get a balance with the reference beam. Another problem is that a reverse gradient must be applied to regenerate the column.

The major advantage of gradient elution is the fact that it provides the maximum resolution per unit time.

When two (or more) solvents are mixed, certain experimental conditions must be met for a successful separation using the technique of high performance liquid chromatography. To maintain the calibration of the detector and to ensure reproducible results, the solvent flow must be controlled accurately. Solvent flows may change because of differences in viscosity and compressibility, and changes in the partial molar volumes of solvents on mixing.

The basic requirements of a gradient elution system are: (1) the gradient must be reproducible and consistent regardless of the solvents used; (2) the flow rate through the column must be constant, since not only will the flow rate affect the retention time but it will also affect quantitative results because it affects peak areas.

In order to get reproducible results, the factors that affect the retention time must be controlled, i.e. column equilibration, temperature, flow rate, and mobile phase. Since in gradient elution the solvent strength is constantly changing during the separation, the process fundamentally involves non-equilibrium. Temperature and flow rate have a relatively minor effect on reproducibility; a $1°C$ change in temperature produces a change in retention time of about 2%, and a 1% change in flow rate produces a 1% change in retention.

The main contribution to changes in retention time comes from the mobile phase composition, and lack of control of the gradient programme is the greatest source of error in the reproducibility sometimes experienced in the gradient elution technique.

References

1. Snyder, L. R. and Kirkland, J. J. (1974) *Modern Liquid Chromatography*, Wiley, New York.
2. Sinsell, J. A., La Rue, B. M. and McGraw, L. D. (1975) *Anal. Chem.*, **47**, 1987.
3. Trappe, W. (1940) *Biochem. Z.*, **305**, 50.
4. Strain, A. H. (1942) *Chromatographic Adsorption Analysis*, Interscience, New York.
5. Jacques, J. and Mathieu, J. P. (1946) *Bull. Soc. Chim. France*, **9**, 4.
6. Snyder, L. R. (1968) *Principles of Adsorption Chromatography*, Marcel Dekker, New York.
7. Snyder, L. R. (1971) *J. Chromatog.*, **63**, 15.
8. Hais, J. M. and Macek, K. (1963) *Paper Chromatography*, Academic Press, New York.
9. Hildebrand, J. H. and Scott, R. L. (1964) *The Solubility of Non-electrolytes*, 3rd Edn., Dover, New York.
10. Keller, R. A., Karger, B. L. and Snyder, L. R. (1971) *Gas Chromatography 1970*, 125.
11. Solvent Use Index Charts, Waters Associates.
12. Snyder, L. R. (1974) *J. Chromatog.*, **92**, 223.
13. Rohrschneider, L. (1973) *Anal. Chem.*, **45**, 1241.
14. Lurie, I. S. (1980) *International Laboratory*, Dec., p. 61.

6

Developing a chromatogram

In this chapter we shall discuss the approach to establish the best analytical conditions for a specific sample. However, most problems can be solved by more than one mode of liquid chromatography, and the final approach may depend on the availability of equipment, columns, or the personal preference of the chromatographer.

6.1 Nature of the problem

It is rare for the analyst to receive an 'unknown' sample; its origin, the use to which it is put, or even its physical form are important indications as to its nature. The more that is known about the sample the easier it will be to decide on the best approach for its analysis. The solubility characteristics of the sample must at least be known since it must be in solution to be injected. The nature of the solvent to be used will also affect the choice of stationary phase. A preliminary analysis by another technique may be necessary; e.g. analysis by i.r. or u.v. spectroscopy would indicate the nature of functional groups present, and a knowledge of the molecular structure will assist the analyst in his choice of detectors.

It is also important to know the nature of the analysis required. Is there only a limited number of components of interest in a multicomponent mixture or is an analysis required of each component present? Would a 'fingerprint' chromatogram of the sample be sufficient or are the components to be separated and collected for further analysis? Is the analysis a one-off problem or will it become a routine method for quality control? All these are questions the answers to which may affect the analyst's approach.

6.2 Choice of Chromatographic Mode

Having determined the extent of the analysis required and the nature of the sample, the analyst is in a position to select the chromatographic mode that is most likely to produce the desired results.

A satisfactory choice of chromatographic mode requires an understanding of the mechanisms controlling retention in the different modes. These have been discussed in Section 1.3 but it is useful to summarize the main points again.

Partition. Stationary phases are invariably of the chemically bonded type, i.e. the 'liquid' stationary phase is chemically bonded to the surface of a base particle, usually silica. Choice of the bonded liquid results in phases of varying polarity. The mechanism of retention is complex but involves some form of partition between the stationary and mobile phases. The retention properties are similar to adsorption under certain conditions and to partition under others. Because of the wide variety of packing polarities and mobile phases which can be used with them, partition chromatography is the preferred choice in the majority of separations.

Adsorption. Separation is the consequence of polar interactions between the active groups on the stationary phase and polar or polarizable functional groups on the solute molecules. The geometric arrangement and number of surface active groups determines the selectivity and the mode is best suited for separation into compound type (e.g. alcohols and esters) and for geometric isomers.

Ion-exchange. Since this involves the substitution of one ionic group by another it is only suitable for ionic or ionizable compounds. The mobile phase is almost exclusively water and pH control is a major factor in selectivity.

Ion-pairing. According to the nature of the system the predominant mode may either be partition or ion exchange. It is thus applicable to non-ionic and ionic or ionizable molecules.

Exclusion. Solutes are separated by differences in molecular size and shape. It is most useful for high molecular weight materials (> 2000) and molecular weight differences of about 10% are required for separation. The mobile phase adopts a non-interactive role in exclusion chromatography.

Bearing in mind the foregoing summary it should be possible to select the mode most likely to lead to a successful separation. Figure 6.1 illustrates how an initial choice may be made on the criteria of molecular weight, solubility and functional group character.

6.2.1 Molecular Weight

With an unknown sample a useful start is to determine its molecular weight range using exclusion chromatography; the resulting size fractions can then be subjected to further exclusion chromatography or they can be analysed by a different chromatographic mode.

The choice of a molecular weight of 2000 is somewhat arbitrary since columns are available for molecular weight separations in the range 2×10^2 to 2×10^6,

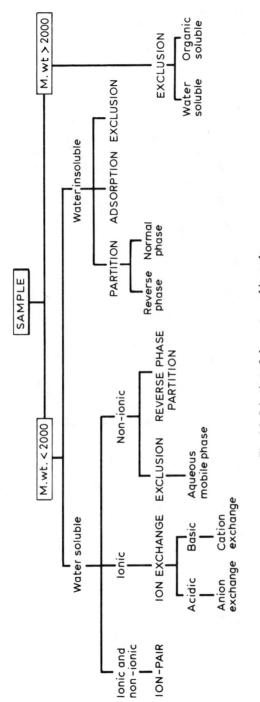

Fig. 6.1 Selection of chromatographic mode.

but generally speaking lower molecular weight materials are better separated by means other than exclusion chromatography. More important is the lipophilic/ hydrophilic character of the sample since some exclusion packings are not suitable for use in aqueous mobile phases whilst others are not suitable for organic phases (see for example Table 4.12).

6.2.2 Solubility

Sample solubility is perhaps the most important guide to the selection of the best mode and column. Thus lipophilic molecules can be separated by reverse-phase partition chromatography on a hydrocarbonaceous phase or by adsorption chromatography. For more polar molecules, soluble in methanol, there is a choice of both normal and reverse-phase partition chromatography though they may also be separated on adsorption packings if the mobile phase is sufficiently polar (e.g. alcohols or chlorinated hydrocarbons). Very polar water soluble molecules can be separated by reverse-partition, ion-exchange or ion-pair chromatography.

6.2.3 Functional Group Character

The presence of ionic or ionizable groups on the molecule indicates that ion-exchange or ion-pair chromatography should be used. Strong hydrogen-bonding functional groups, e.g. alcohol or amine, are particularly amenable to adsorption chromatography especially where the molecules differ in the number and configuration of the hydrogen-bonding groups whereas molecules which are predominantly aliphatic or aromatic where solubility differences can be exploited will best be separated by partition chromatography.

A correlation between functional group characteristics and chromatographic mode is shown in Fig. 6.2. The choice between adsorption and a bonded phase packing may often be one of convenience and availability.

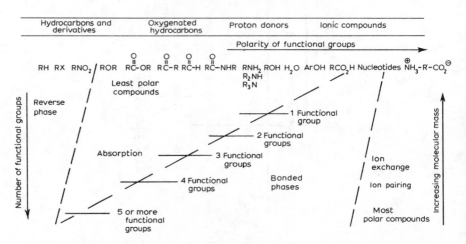

Fig. 6.2 Correlation between packing and sample type (after Waters Associates).

6.3 Selection of Stationary Phase and Mobile Phase

A successful separation is achieved when a proper balance is established between the intermolecular forces involving the sample, the mobile phase (solvent), and the stationary phase. The intermolecular forces may be measured in terms of the polarity of the molecules, and most good separations are achieved by matching the polarity of the sample and stationary phase and using a mobile phase of different polarity. If the sample is too similar to the mobile phase in terms of polarity, the stationary phase is unable to compete successfully for the sample and there is little retention. In that case the polarity of the mobile phase or of the stationary phase or of both must be changed. When the stationary phase is more polar than the mobile phase the system is referred to as 'normal phase liquid chromatography'. In some cases the sample is so strongly retained by the stationary phase that even substantial changes in mobile phase polarity do not decrease the retention time sufficiently. In this situation it is appropriate to use 'reverse phase chromatography' in which the stationary phase is less polar than the mobile phase.

Fig. 6.3 shows the general interactions between sample and mobile phase as a function of polarity. Thus in *normal* phase chromatography the *least* polar component is eluted first and *increasing* the mobile phase polarity *decreases* the elution time, whereas in *reverse* phase chromatography the *most* polar component is eluted first and *increasing* the mobile phase polarity *increases* the elution time.

6.3.1 Stationary Phase

The general nature of the stationary phase will be determined by the selected mode for the separation. But within a given mode there is still a considerable choice of stationary phase to be made: e.g. which molecular weight range exclusion packing to use; whether to use a weak or strong anion or cation exchange packing; the polarity of an adsorbent. Because of the complex nature of the partition process the most difficult choice is when using chemically bonded partition packings, but this complexity also accounts for the variety of separations which may be achieved on these packings. The packings available range from those with polar functional groups $-OH$, $-NH_2$, $-CN$) to the apolar hydrocarbon types. The $-CN$ type can be used to separate many of the polar compounds which may be separated by adsorption chromatography on silica. However the bonded phase is more suitable for gradient elution separations. The NH_2 phase can operate in both the normal and reverse phase modes as well as a weak ion exchanger. Thus it can be used to separate polar compounds e.g. substituted anilines, esters and chlorinated pesticides in the normal phase mode, carbohydrates in the reversed phase mode and organic acids (e.g. dicarboxylic acids) in the ion-exchange mode. Because of long equilibrium times, due to the polarity of the amino group, it is not recommended that a single column is used in all three modes but that a separate column should be kept for each mode.

The simplest criteria for the selection of the stationary phase is that 'like has an affinity for like'. If the sample is soluble in hydrocarbon solvents, indicating

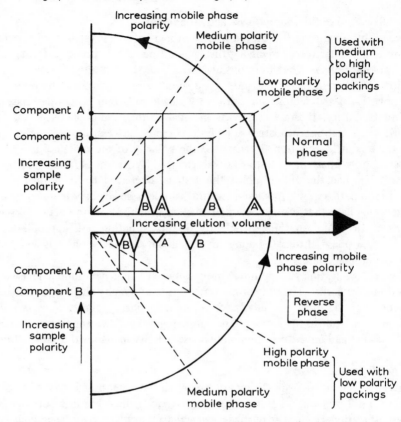

Fig. 6.3 Interactions between sample and mobile phase as a function of polarity
(after Waters Associates).

its own low polarity, then an apolar stationary phase should be tried first. Conversely if the sample is soluble in polar solvents such as water, indicating a high polarity sample, then a polar stationary phase should be tried. This approach to column selection in partition chromatography using solubility in a hydrocarbon solvent, in alcohol and in water is exemplified in Table 6.1 which also lists typical mobile phases and gives examples of sample type.

6.3.2 *Mobile Phase*
In liquid chromatography the mobile phase may be any single liquid or combination of liquids which are compatible with the sample, column and instrumentation. Physical characteristics, however, such as viscosity, volatility, compressibility, refractive index and u.v. absorption can limit the choice of mobile phases.

Table 6.1 *Column selection for chemically bonded phases*

Sample type	Mode	Column type	Mobile phases	Examples
Low polarity-soluble in hydrocarbons	Reversed phase	$-C_{18}$	methanol/water, acetonitrile/water acetonitrile/tetrahydrofuran tetrahydrofuran/methylene chloride	polynuclear aromatic hydrocarbons, triglycerides, lipids, esters, fat soluble vitamins, steroids, hydroquinones, sulphamides, alkaloids
Medium polarity-soluble in alcohol	Normal phase	$-CN$	acetonitrile, methylene, chlorides, hexane chloroform.	fat soluble vitamins, steroids, aromatic alcohols, amines, esters, lipids (class separation) analgesics
		$-NH_2$	hexane, methylene chloride, isopropanol.	aromatic amines, esters, chlorinated pesticides, carboxylic acids lipids (homologue separation) nucleotides, phthalic acids, poly-nuclear aromatics
	Reverse phase	$-ODS$ $-C_8$ $-CN$ $-TMS$ $-NH_2$	methanol, water, acetonitrile	steroids, alcohol soluble natural products
				vitamins, aromatic acids, xanthines antibiotics, anticonvulsants, food additives carbohydrates
High polarity soluble in water	Reversed phase	$-C_8$ $-CN$ $-TMS$	methanol, acetonitrile water, buffered water.	water soluble vitamins, amines, aromatic alcohols, analgesics, antibiotics, food additives
	Cation exchange	$-SO_3^-$	water & aqueous buffers	inorganic cations, carbamates, vitamins, amino acids, catecholamines, nucleosides, glycosides
	Anion exchange	$-NR_3^+$ $-NH_2$	citrate, phosphate buffers pH 2–7	nucleotides, inorganic anions, sugars organic acids, analgesics
	Ion pair (reverse phase)	$-ODS$, C_8 C_2	water, methanol, acetonitrile and counter ions.	acids, sulphonated dyes, catecholamines

The role of the mobile phase may either be interactive, if a change in mobile phase composition produces a corresponding change in retention, or non-interactive if the retention is independent of mobile phase composition. Thus in adsorption, partition and ion-exchange the mobile phase is interactive, whereas in exclusion chromatography it is non-interactive. The ability to change the selectivity of the column by altering the mobile phase composition is a powerful variable and has made it possible to reduce the number of stationary phases necessary in liquid chromatography. However, finding the best mobile phase system may lead to lengthy method development and it is in this area that fully programmable microprocessor controlled chromatographs can be invaluable.

Mobile phase (solvent) strengths are usually measured in terms of the solvent polarity and are presented as the eluotropic series. A more fundamental measure of solvent polarity is defined by the Hildebrand solubility parameter δ, and for normal phase chromatography the strength of the mobile phase increases with increasing δ values. Solvents in the eluotropic series with comparable polarity rankings may have different solubility parameters. However, the eluotropic series is a useful means of ranking the relative polarities of solvents, and it is in common use. Fig. 6.4 shows the eluotropic series for many of the common solvents used in HPLC. Together with these are shown examples of various types of stationary phase. The full lines represent the range of solvents most frequently used with these phases.

6.4 Choice of Detector

The choice of detectors is still strictly limited. The refractometer is the only universal detector readily available but it suffers from lack of sensitivity and the inability to cope with solvent programming other than of a stepwise nature. A variable wavelength u.v. detector offers the best choice for a large range of solutes but in specific cases the fluorescence or electrochemical detector can be used. Unless a solute system is reasonably well defined, whichever detector system is used, there will be a chance that solutes will not be detected, and a combination of detectors may be necessary.

6.5 Chromatographic Separation

Having considered the characteristics of the sample and decided on the mode of separation, the stationary phase, and the mobile phase, the analyst is now ready to make his first injection. An analytical column (4–5 mm i.d. x 25 cm long) at room temperature should be used for this purpose.

Before an injection is made the sample must be dissolved in a suitable sample solvent. Ideally the sample solvent should be the same as the mobile phase, but if the sample is not sufficiently soluble in the mobile phase another solvent can be used. This sample solvent must then be of high purity so that extraneous peaks are not introduced, be miscible with the mobile phase and should give a minimal detector response, otherwise peaks eluting just after the sample solvent may be

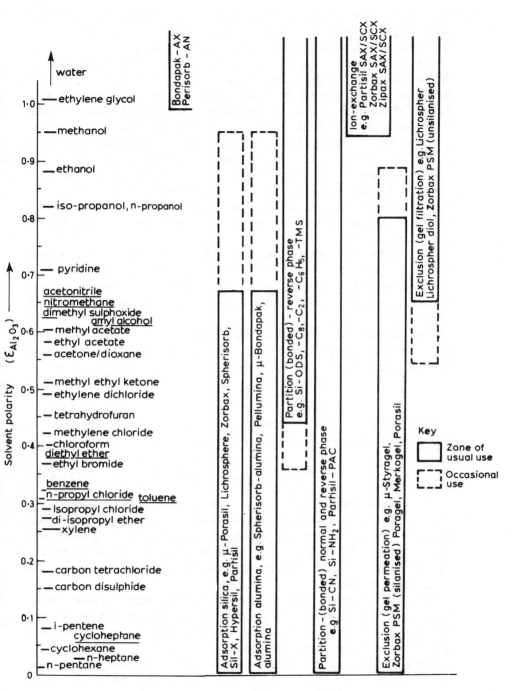

Fig. 6.4 Elutropic series and column packings.

masked. Complex samples, e.g. body fluids, should be filtered or centrifuged before injection to remove particulate matter which may interfere with the proper functioning of pump check valves or block transfer lines.

The chromatographic process must maintain the integrity of the sample, i.e. decomposition and/or degradation of the sample or irreversible adsorption of the sample must be avoided. Spurious or ghost peaks may be generated during a gradient elution when the mobile phase strength becomes sufficient to elute a previously retained sample component. It is therefore good practice to run a blank chromatogram before injecting the sample. The sample solvent, too, should be injected to check it for impurities. The initial chromatogram is unlikely to be completely successful; there may not be any resolution of the components, or some may be only poorly resolved, and the analysis time may be too long. However, by applying the simple theory presented in Chapter 2, the analyst can rapidly select the experimental conditions necessary to achieve the maximum separation in the minimum time.

In considering the optimization of conditions for analysis, although the same general principles are applicable to all modes of chromatography, the means of achieving the desired result may be different. It will be best, therefore, to discuss the general principles and then to discuss them in a consideration of the different chromatographic modes.

Two chromatographic parameters are important in the optimization to give maximum separation in the minimum time. Equation (2.20) relates the retention time (t_R) to the capacity factor (k') thus:

$$t_R = \frac{L}{v_m} (1+k') \tag{2.20}$$

and Equation (2.58) relates the resolution R_S to the capacity factor (k'), relative retention (α) and the number of theoretical plates (N) thus:

$$R_s = \frac{1}{4} \left(\frac{\alpha-1}{\alpha}\right) \left(\frac{k'}{1+k'}\right) N^{\frac{1}{2}} \tag{2.58}$$

As discussed in Section 2.10 the optimum value of k' lies between 1 and 10. Larger values of k' lead to longer retention times, and the chromatographic bands are eluted as wide, flat peaks which are difficult to detect. If k' lies in the optimum range, therefore, it is undesirable to change its value, and changes in resolution can be brought about only by increasing the number of theoretical plates (N) or the relative retention (α).

The capacity factor is the parameter which is most easily optimized since it usually only involves a change in the mobile phase strength. However, k' values may also be adjusted by a change in stationary phase polarity, although this approach is obviously not so convenient.

The optimum value of α lies between 1.05 and 10. A change in α is achieved by altering the nature of the stationary phase and/or the mobile phase. However in changing the mobile phase it is the composition rather than the solvent strength which is the more important. The effect of changes in α are more difficult to predict than are changes in k' and N, and in a multicomponent mixture changes in α may merely result in a reshuffling of the elution order. In terms of analysis time, however, a change in α is often the best way of improving R_S.

Since it is easier to achieve an increase in R_S by increasing N, this may be preferable. However, increasing R_S in this way generally results in longer retention times, since an increase in N is usually achieved by an increase in column length or a reduction in mobile phase flow rate. If a routine analytical method is being established it may be worth the extra effort to change α to achieve the required resolution, there being a saving in time on subsequent analyses.

An increase in N results in less band broadening and greater resolution, and can always be achieved by increasing column length, increasing the pressure drop across the column, or increasing the analysis time. Unlike increases in k', N can be increased in theory without limit; however, there are practical limits set by the available pressure drop and also the packing characteristics of the stationary phase.

The value of N for an LC column is dependent on a large number of variables. Some of these (choice of mobile phase and stationary phase) may already have been selected to optimize k' and α values. Others (particle size and temperature) can be optimized independently. This leaves the analyst with three interdependent parameters which affect the value of N: column length (L), pressure drop (ΔP), and analysis time (t). These parameters can be varied to increase N in three ways: (a) decrease ΔP holding L constant; (b) increase L holding ΔP constant; (c) increase L and ΔP holding t constant.

Method (a) is obviously the simplest although it will result in longer analysis times. Method (b) may be preferred when a large number of analyses are to be performed; although time may be spent on preparing longer columns, the analysis time per sample will be less. Method (c) is only an option if the LC unit is already operating below its maximum pressure, or if a pumping unit with a higher maximum pressure is available.

Although temperature may not have dramatic effects it can be significant. An increase in temperature will give an increase in N by improving mass transfer; particularly in ion-exchange processes, and will also give smaller k' values; the effect on α is difficult to predict.

To a large extent then optimization of the separation is brought about by changes in the mobile phase, the stationary phase having been selected in the first place to provide a suitable chromatographic mode for the analysis. This is not to imply that there is no choice, for in partition chomatography the choice is considerable.

6.5.1 Adsorption
The choice of adsorption packings is limited to silica and alumina with alumina having a greater selectivity toward unsaturated compounds. Since both of these

are strongly polar the problem is usually that of too high k' values rather than too small. Since they are both used in the normal phase mode, k' values may be decreased by using a more polar mobile phase. k' values can be increased by using a packing with a larger surface area. Most of the spherical silicas have surface areas $\sim$ 200–250 m²g⁻¹, but the irregular ones tend to have higher values (400–500 m²g⁻¹). In order to get reproducible k' values the water content of the mobile phase must be carefully controlled. This is most easily done by saturating the mobile phase with water.

Selectivity may vary slightly with the pore diameter, but the use of a secondary solvent produces more predictable selectivity changes.

6.5.2 Partition

It is in the partition mode that the chomatographer has the widest choice of stationary phases and if conventional coated partition packings are included the choice is wider still. However, literature applications are almost exclusively with chemically bonded packings, which may be divided into those containing polar functional groups and the non-polar packings.

Compared to silica the polar bonded phases respond more rapidly to mobile phase changes and show less peak tailing since the highly polar silanol groups are replaced by somewhat less polar groups. The weakly polar packings include diol, dimethylamino or nitro groups. The diol is useful for very polar compounds, e.g. organic acids, whilst the nitro group shows a greater selectivity toward aromatic molecules. The dimethylamino function can also act as a weak anion exchanger.

Moderately polar packings all contain a cyano function, either alone or as a cyanopropyl or cyano-amino grouping. The CN packings are an alternative to silica but they give lower k' values, although the selectivity is about the same. It is a very useful general purpose packing usually used in the normal phase mode.

The cyano-amino packing (Whatman. Partisil PAC) can be used in both normal and reverse phase mode. The secondary amine group shows a different selectivity to a primary amine and since it can also be protonated it can function as a weak anion exchanger. Its main application is for highly polar compounds where it is comparable in selectivity to conventional 'ether' and 'oxynitrile' phases.

The most polar packings contain amino or aminopropyl functional groups. Since they are basic they give rise to quite different selectivities to silica when used in the normal phase mode for the separation of polar compounds. In the reverse phase mode they can be used to separate mono, di, tri and polysaccharides using acetonitrile: water (80:20). They can also be used as anion exchangers for organic acids (e.g. carboxylic acids).

The non-polar bonded phases used in a reverse-phase mode are the most frequently used packings. They are able to provide separation of non-ionic, ionic and ionizable solutes by the use of suitable mobile phases and the retention is predictable since the k' values usually decrease as the hydrophilic character of the solute increases. Column equilibration with mobile phase is very rapid so that gradient elution causes no problems.

All the non-polar bonded phases contain a hydrocarbon or phenyl function. However, the method of preparation of the bonded phase can have a profound affect on the resulting properties. The object in preparing a bonded phase is to cover up as many of the silanol groups as possible. This is best achieved by using monofunctional or bifunctional silanes (Fig. 6.5).

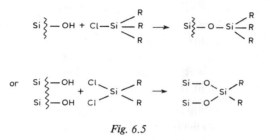

Fig. 6.5

The use of trifunctional groups or the presence of water may lead to the formation of a polysiloxane polymer. This polymer network leads to poor mass transfer and, therefore, lower efficiencies. The polymerization reaction is difficult to control and, therefore, batch-to-batch reproducibility is also not so good. However, such polymeric phases are suitable for the separation of very non-polar solutes [1]. Since the silica backbone is better protected these columns are also more stable under conditions of high pH.

A typical silica contains about $7\,\mu$mol silanol groups per square metre of surface of which approximately $4\,\mu$mol m^{-2} can react. Thus even with the highest surface coverage possible there are still $\sim 3\,\mu$mol m^{-2} silanol groups unreacted. These OH groups could be responsible for adsorption affects giving rise to mixed mechanisms, but the exact nature of the role of such residual groups is not clear. Some manufacturers 'cap' the excess silanol groups by reacting with a small silane $(CH_3)_3\,SiCl$ to reduce the possibility of mixed mechanisms. Because of the complexity of the bonding process there is a considerable difference between bonded phases from different suppliers and a separation achieved on one column may not be reprodicible on a similar column but from a different manufacturer.

The most commonly used reverse phase packing is that containing the octadecyl group C_{18} (ODS). They are available with carbon loadings from 5–40%. The lower the carbon loading the more likely it is that residual silanol groups will be present.

The monomeric and low-loaded ODS columns are best for non-polar species and for ion-pair chromatography whereas the polymeric and heavily loaded columns are only suitable for the separation of non-polar species, since more polar solutes show poor mass transfer in the polymeric phase.

The octylsilanes (C_8) and other short chain packings are better for more polar solutes; this is probably due to an increase in availability of surface silanol groups which since they are hydrated allow the polar molecules to 'wet' the surface. They also give lower k' values.

The C_2 phase may give mixed retention mechanisms and column stability may be reduced and the phenyl phases are recommended for more polar compounds, e.g. fatty acids and peptides.

In general it is found that:

1. the retention is proportional to the carbon chain length and to the carbon loading.
2. a higher carbon loading gives more stable columns.
3. both alkyl chain length and carbon loading may affect selectivity—a long chain length and higher carbon loading have a higher selectivity for larger non-polar molecules than for small solute molecules.
4. the role of accessible Si—OH and Si—O—Si groups is not clear, but they are likely to lead to mixed mechanisms of retention.

The effect of mobile phase strength on k' values is shown in Fig. 6.3. Using the polar bonded phases in the normal phase mode a wide choice of mobile phases is available. Increasing the solvent strength (i.e. increasing the proportion of the more polar solvent) decreases the k' values. In the reverse phase mode the normal solvents are water, methanol and acetonitrile. In this case an increase in solvent strength means an increase in the proportion of the less polar solvent and gives a decrease in k' values; hence k' values will change predictably with a change in solvent strength.

The control of pH is another way that k' values may be predictably changed. If the solute is ionisable, e.g. an aromatic acid, a low pH buffer will suppress the ionization making the solute less hydrophilic and will increase the k' value in a reverse phase packing. Using a high pH buffer (NaOAc) the ionized form will be favoured and k' would decrease.

To reduce peak tailing and to increase the efficiency of the separation small amounts of modifiers are often added to the mobile phase, e.g. sodium phosphate or sodium acetate in the reverse phase mode and acetic acid in the normal phase mode. Using mobile phases containing methanol or acetonitrile, 1–2% tetrahydrofuran may have the same affect.

With the increasing availability of chromatographs with ternary mobile phase capabilities the optimization of the mobile phase for selectivity is becoming increasingly complex. The use of a solvent characterization parameter P' to distinguish between solvent strength and solvent selectivity, as proposed by Snyder, has been described in Section 5.4. This approach has been combined with a mixture-design statistical technique to optimize solvent strength and solvent selectivity for reversed phase liquid chromatography and a new method of measuring the quality of a separation called overlapping resolution mapping (ORM) has been introduced [2]. The technique is claimed to predict the best mobile phase composition to give a specified minimum resolution of all pairs of peaks. The interested reader is referred to reference [2].

6.5.3 Ion-Pair Chromatography

Ion-pair chromatography is a method of controlling both retention and selectivity by the addition of an ion-pairing agent to the mobile phase and can be used for samples containing both ionic and non-ionic components.

The technique is usually used in the reverse-phase mode using an ODS stationary phase. The length of the alkyl chain may or may not be significant since the mechanism may vary with the chain length. The usual solvents therefore are methanol/water or acetonitrile/water. The main limitation on the choice of mobile phase is the solubility of the ion-pairing agent. When using acetonitrile the solubility may be increased by using 90/10: ACN/H_2O as the organic modifier. If the ion-pairing agent is of low solubility it may precipitate out and cause plugging of tubing.

The ion-pairing agent is usually a large ionic or ionisable organic molecule, although perchloric acid ($HClO_4$) is also used. Typical ion-pairing agents and their applications are shown in Table 6.2.

Table 6.2   *Ion-pairing agents*

Type	Application
Quaternary amines e.g. $(CH_3)_4N^+$, $(Butyl)_4N^+$	strong or weak acids; sulphonated dyes, carboxylic acids
Tertiary amines e.g. trioctylamine	sulphonates
Alkyl and aryl sulphonates e.g. camphor sulphonic acid	strong & weak bases; benzalkonium salts, catecholamines
perchloric acid	for a wide range of basic solutes e.g. amines
alkylsulphates e.g. lauryl sulphate	similar to sulphonates but show different selectivity

Although the mechanism of ion-pairing is subject to debate several of the factors which affect resolution have been defined.

Effect of counter-ion. It is found that retention is proportional to the ability of the counter ion to ion pair, the size of the counter-ion and, within limits, to its concentration.

Nature of the stationary phase. In general the k' values are increased by using a stationary phase with a longer hydrocarbon chain length (i.e. more lipophilic) with a higher carbon loading. Hence ODS is usually used.

Addition of an organic modifier to the mobile phase. If k' values are too high the addition of an organic modifier to the mobile phase will reduce k'. This is because the ion-pair and the modifier will be in competition for the adsorption site. Consequently, the more lipophilic the modifier the greater will be the reduction in k'.

Control of pH. The pH of the mobile phase is usually used to give the maximum concentration of the ionic form. However, it may also be used to suppress ionization and to reduce k', for example if the solute is a strong acid ($pK_a < 2$) or a strong base ($pK_b > 8$), the solute will be completely ionized in the usual working pH range of 2–8. However, the pH may still be adjusted in this range to give selectivity if weak acids or weak bases are also present. Weak acids will be ionized in the pH range 6–7.4 and their k' values will be dependent on the ion-pair formed. Below pH 6 they will not be ionized and their retention will depend on their lipophilic character. Similarly for weak bases; these will be ionized below pH 6 but will undergo ion suppression above this pH.

Effect of Temperature. The retention will be expected to decrease as the temperature increases and there may also be an increase in efficiency.

6.5.4 Ion-Exchange

The choice of the type of ion-exchange column to be used depends on the characteristics of the solutes to be separated. Thus, whereas a strong anion exchanger will absorb weak and strong anionic species whether acidic or basic, a weak anion exchanger will not absorb anions from the salts of strong bases. A strong cationic exchanger will absorb acidic and basic cationic species, but a weak cationic exchanger will not absorb cations from the salts of strong acids.

The most important single parameter in ion-exchange is pH. For example, to be retained on an ion-exchange packing the sample must be ionized. If the pH of the solution equals the pK_a (or pK_b) of the acid (or base) the molecules are in equilibrium with their ions and are 50% dissociated. To ensure complete dissociation, if the pK_a (or pK_b) is known, a rule of thumb for the starting pH is to take the highest pK_a value of the molecules in the sample and add 1.5. For bases, take the lowest pK_b and subtract 1.5.

In the first instance, however, the pH range to be used in an ion-exchange separation is determined by the type and nature of the stationary phase. Thus those based on a silica bonded phase are limited by the stability of the silica itself and cannot be used with a pH much higher than 7.5–8.0. Coated pellicular packings are less susceptible to alkaline attack and can be used up to a pH of 9.0. Packings based on a styrene–divinylbenzene resin have much wider pH ranges; 0–12 for an anion exchanger and 0–14 for a cation exchanger.

The ability of an ion-exchanger to act as such is expressed as the exchange capacity and this depends upon the degree of ionization of the resin. At low pH strongly acidic cation exchangers are neutralized by H^+ ions and its exchange capacity falls rapidly below a pH of 2. For a weakly acidic cation exchanger this fall off in capacity occurs below about pH 6. Similarly, at high pH anion exchanges are neutralized by OH⁻ ions. For strong anion exchangers this occurs above pH 11 and for weak anion exchangers above pH 8. Strong ion exchangers may therefore be used over a wider range of pH than can weak anion exchangers; hence the popularity of the sulphonate group for cation exchangers and the trialkylammonium group for anion exchange in HPLC ion exchange packings. Thus k' values may be decreased by changing from a strong to a weak ion exchanger or by varying the pH to reduce the ionization of the packing.

The majority of separations by ion-exchange are brought about by changes involving the mobile phase. The mobile phase may be considered to be water buffered with various salts (e.g. phosphate, borate, citrate, acetate) to which may be added water-miscible organics (e.g. methanol, acetonitrile). Solvent strength and selectivity depend on the nature and concentration of buffer and other ions present and of any organic additives and on the pH. Whereas the effect on retention is reasonably predictable changes in α are not. The simple picture of ion-exchange may be supplemented by partitioning, complexation and ligand-exchange which make it difficult to predict the effect of any changes.

The ionic strength of the mobile phase is proportional to the concentration of the ions in solution and to the valency of those ions. Hence, ionic strength can be varied by changing the concentration of the buffer (keeping pH constant) or by adding an addition salt. The usual range of ionic strength is from 0.001–2 mol dm^{-3}. In either case the solvent strength increases as the ionic strength increases and leads to a reduction in retention because there is more competition for the ion-exchange sites. Provided the sample ions all have the same valency the change in ionic strength will affect all sample ions equally, so that selectivity is not changed. The use of divalent ions causes even larger changes in k' since $k' \propto [1/(y\pm)^2]$ instead of $k' \propto [1/(y\pm)]$ for a univalent ion.

The nature of the mobile phase counter-ions also affects the solvent strength. In anion exchange the retention of a given sample generally increases in the sequence:

$$\text{citrate} < SO_4^{2-} < \text{oxalate} < I^- < NO_3^- < Br^-$$
$$< SCN^- < Cl^- < \text{formate} < \text{acetate} < OH^- < F^-;$$

and for cation-exchangers:

$$Ba^{2+} < Sr^{2+} < Ca^{2+} < Ni^{2+} < Cd^{2+} < Cu^{2+} < Co^{2+} < Zn^{2+}$$
$$< Ag^+ < Rb^+ < K^+ < NH_4^+ < Na^+ < H^+ < Li^+.$$

However, there may be considerable variation in these series so that they are best regarded as approximate. Although the major effect is on retention selectivity may also change.

The most commonly used factor for control of retention is the mobile phase pH. In cation exchange the more basic molecules would elute with the higher k' values. An increase in pH would lead to a decrease in ionization and to lower k' values. In anion exchange the more acidic molecules would elute with higher k' values and decrease in pH would lead to a decrease in ionization and a decrease in k' values. Thus in cation exchange an increase in pH is equivalent to an increase in solvent strength and in anion exchange a decrease in pH is equivalent to an increase in solvent strength. In both cases changes in selectivity may also occur especially in anion-exchange where pH is by far the most powerful mobile phase property to change.

The effect of organic modifiers in the mobile phase is similar to the addition of an organic modifier to water in reversed-phase chromatography. Thus the retention is decreased as the amount of organic modifier is increased and the result is greater the more non-polar the solvents; alcohols and glycols are most commonly used as modifiers.

Ion exchange gives lower column efficiencies than other forms of liquid chromatography. If sample and column stability allow, higher temperatures can be used to advantage to improve efficiency and hence resolution. Retention will decrease as the temperature increases and selectivities may also vary especially if the compounds being separated have different valencies and/or structures.

Changes in α are brought about by several factors, and since there may be several competing equilibria within any given ion exchange system it is difficult to predict the effect of these changes.

It will be sufficient to list the changes that affect α:

1. a change in the pH of the mobile phase will bring about a change in α as well as in k' as described previously;
2. variations in the hydrated ionic radius of the sample;
3. reduction of the common or competing ion in the mobile phase;
4. use of a different packing with comparable ionization properties;
5. temperature changes may produce a change in α since the change may affect different exchange reactions to a different degree;
6. the addition of organic molecules to the mobile phase may change α by changing solubilities, dissociation constants, the ionic radius of the sample, and rates of exchange by forming ion-pairs or complex molecules.

This latter possibility is particularly prevalent when buffers such as EDTA, citric acid, or a polyvalent salt are used.

Although the complexity of the system makes it difficult to predict the effect of a change on the resolution, it is the variety and subtlety of these changes that makes ion-exchange such a useful separative technique.

6.5.5 Exclusion Mode

The factor controlling retention in 'pure' exclusion chromatography is the pore size of the gel. Since the mobile phase is non-interactive it will have no affect on the retention, but it can affect the resolution. Retention can be increased by using a gel with a higher exclusion limit (i.e. a larger pore size) and decreased by using one with a lower exclusion limit (smaller pore size).

The mobile phase is selected for its ability to dissolve the sample. However, viscosity is also important as a low mobile phase viscosity will lead to high N values and an improvement in resolution. An increase in temperature will achieve a similar result. The mobile phase must also be compatible with the column. In aqueous systems the mobile phase pH must be in the range 2–8.5 if silica packings are used and very polar mobile phases are not used if polystyrene packings are used.

In 'pure' exclusion chromatography there should be no interaction between the sample and the stationary phase. Adsorption is particularly prevalent on silica-based packings. This can be prevented by selecting a mobile phase which will have a higher affinity for the packing and will be preferentially adsorbed.

The resolution should first be improved by using the flow rate corresponding to maximum efficiency. An increase in column length will also improve resolution since the degree of separation is directly related to column length. This can either be achieved by adding to the actual column length by using longer columns or coupling columns in series, or by increasing the effective column length by re-cycling the partially resolved sample through the same column until the desired resolution is obtained.

The main advantages of re-cycling are that lower pressures may be used and, since no extra columns are required, costs in terms of column packings and solvents are minimized.

Resolution can also be increased by using a smaller diameter packing with a narrow particle size distribution.

If adequate resolution is still not obtained it means that the molecular size differences are too great to be dealt with on a single column, and two or more columns of varying pore sizes should be connected in series. It is important that the columns are placed in order of increasing pore size so that the sample first enters the column of smallest pore size.

6.6   The General Elution Problem

The general elution problem occurs in all forms of chromatography when the k' values for the first and last components of a multicomponent system differ by more than a factor of about 10, i.e. it is impossible for all the k' values of the components to fall simultaneously in the optimum range $1 \leqslant k' \leqslant 10$. Under these conditions the chromatogram would appear as in Fig. 6.6(a). Sample components 1–4 are poorly resolved because their k' values are too low. Components 5–8 are well resolved but peak 8 is becoming too diffuse and components 9 and 10 are too diffuse for optimum detection. The most common way to overcome this problem is to use the technique of gradient elution whereby the composition of the mobile phase is varied continuously throughout the analysis. Using a weaker solvent, peaks 1–4 would have higher k' values and would show better resolution. As the strength of the solvent is increased the latter peaks have their k' values reduced and the final chromatogram appears as in Fig. 6.6(b). Some of the problems of gradient elution have been discussed in Section 5.7.

In optimizing the separation using gradient elution the type of gradient to be used must first be considered. Linear gradients often give the best results since the peaks are equally spaced and the band widths are approximately equal; further-

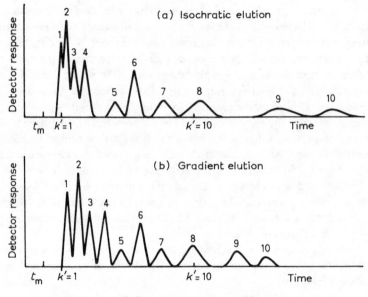

Fig. 6.6 The general elution problem.

more it is often easier to see whether a change in programming rate would be an advantage or not. The use of more complex gradients such as non-linear gradients or linear gradients with constant composition holds require more sophisticated and hence expensive equipment but they do accommodate larger variations in k' values. The number of solvents used in the gradient is usually limited to three. This restricts the possible change in solvent strength since if the polarities of the solvents are too dissimilar they may only be partially miscible (e.g. water and ethyl acetate or diethyl ether). The use of step-wise elution overcomes this problem since the method lends itself to many solvents. This method does require a series of solvent reservoirs each containing a premixed solvent and is therefore less convenient than using continuous gradients. However no special equipment is required.

The choice of mobile phases for gradient elution is only limited by the requirement that they must be miscible and also non-reactive. In normal phase chromatography the mobile phase is usually a hydrocarbon modified by a halogenated compound e.g. n-heptane/dichloromethane and in the reverse phase mode water/methanol and water/acetonitrile are most commonly used.

Quite large viscosity changes may be observed when using water/methanol mixtures, the maximum viscosity occurring at 40% methanol. These viscosity changes manifest themselves as a change in column inlet pressure if a constant displacement pump is being used. Water/acetonitrile mixtures do not show the same large viscosity effects; however acetonitrile is rather toxic. Tetrahydrofuran is another frequently used modifier being less polar than methanol or acetonitrile.

Similar viscosity changes may be observed with the hydrocarbon solvents. Thus pentane or hexane would be preferable to octane but their higher vapour pressures may cause pumping problems. This also applies to dichloromethane which also has a high vapour pressure.

In establishing a programmed gradient run it is necessary to determine the initial conditions, the programme rate and the regeneration sequence. The regeneration sequence is required to return the column to its initial condition ready for the next injection. The application of a reversed programme avoids miscibility problems which could occur if the programme went from final to initial conditions without the intermediate steps.

The following sequence is recommended:

(i) On the first run start with a slow (2% per min) linear gradient commencing with the weakest solvent and increasing solvent strength.

(ii) Note the solvent composition when the first peak is eluted and use 50% of this solvent composition (% solvent B) as the initial solvent composition for the next trial run, e.g. if the first peak eluted at 10% methanol in water use 5% methanol in water as the initial composition. This is because it is found empirically that the solute does not start to move through the column until the solvent composition (in terms of the percentage of the stronger solvent) is half of that which gives a $k' = 1$. If the per cent B is less than 5 it may be necessary to start the programme with 100% A—the weaker solvent.

(iii) Optimize the programme by starting with the low k' solutes. Remember that if the programming rate is increased the analysis time will be reduced but so will the resolution.

(iv) Adjust the flow rate, temperature etc. to give the desired resolution.

(v) Use the reverse programme for column regeneration.

Before carrying out an analysis using gradient elution it is good practise to do a blank run, because insoluble material which had previously been deposited at the top of the column may be eluted when the solvent strength increases and solubilizes this material. These peaks will elute as relatively sharp 'ghost peaks'.

The 'general elution problem' can also be solved by varying the column temperature of the mobile phase flow rate.

Temperature programming, which proved a great benefit in gas chromatography, has not been so useful in liquid chromatography. As the temperature is increased, k' values are decreased and solutes are eluted more quickly. However, the accompanying viscosity changes and the lack of thermal equilibrium in the column reduce resolution, and there is little to be gained by the technique. Programming the flow rate can improve the resolution of the chromatogram, but the required equipment is expensive and gradient elution is to be preferred.

An alternative way to solve the general elution problem is to analyse the sample with two or three different solvent systems. The separation of hydrocarbons, tri-, di-, and mono-glycerides has been achieved by Aitzetmuller [3] using two different gradient programmes with three solvent systems (see Chapter 8): heptane–diethyl ether, chloroform–dioxan, heptane–diethyl ether–dioxan–propan-2-ol–water. However, the same mixture can be analysed by eluting the hydrocarbons and triglycerides with a diethyl ether–hexane (1 : 19) solvent from the solute mixture [4].

A second sample of the solutes can be eluted with diethyl ether–hexane (1 : 3); the hydrocarbons and triglycerides will be eluted together followed by two peaks for 1,2-diglycerides and 1,3-diglycerides.

Finally a third sample can be eluted with diethyl ether–hexane (4 : 1); the hydrocarbons, triglycerides, and diglycerides are eluted together whilst a second peak represents monoglycerides.

The overall composition can be determined by combining the results of the three analyses. Clearly this is a slower procedure than programmed gradient elution, especially since column regeneration is needed after each analysis.

The use of coupled columns has been suggested as another way to solve the general elution problem. A preliminary separation is performed on one column (A) which is linked to two or more additional columns in parallel. The peaks with low k' values eluting first are switched to column B which has a different stationary phase from A. The same solvent carries these early peaks through to column B where their k' values are increased. Components eluting from column A with intermediate polarity are passed to column C which has a third different stationary phase and improves the separation for these solutes. Finally, the most polar solutes are eluted last from column A and passed to column D which has a fourth stationary phase permitting these components to be eluted with better separation. The major advantages of this technique are that resolution per unit time is as good as gradient elution and that column regeneration is *not* necessary.

6.7 Quantitative Analysis

Quantitation in LC is no different from any quantitative analytical method in the sense that good analytical practise is required. This means that due attention must be paid to the following points:

1. what type of analysis is required; is it a single or multicomponent analysis?
2. what are the concentration levels to be measured?
3. are pure standards available for calibration?
4. what interferences might be expected from the sample matrix?
5. what accuracy and precision is required?

For a multicomponent analysis it is essential that good resolution (preferably baseline $R_s = 1.5$) for all components is obtained.

A knowledge of the concentration levels to be expected is also necessary since different levels have their own problems. For example, in trace analysis cross-contamination of samples is a potential source of error. Noise and drift caused by solvent contaminants, flow variations and temperature fluctuations may decrease the precision. Purity determinations in the 95–100 per cent range are often difficult to reproduce.

Pure standards are required for both qualitative identification and quantitative calibration of the detector. Detector response may vary markedly from solute to solute so that calibration is much more important than in GC.

Matrix effects may give spurious results from unwanted or unknown impurities and often the separation of pure samples varies considerably from the actual sample because of these effects. Hence where possible a 'synthetic' sample as near to the actual sample in terms of composition and concentration should be studied to establish any matrix effects.

Accuracy of an analysis refers to the nearness of the measured value to the true value. This is dependent on calibrating the system with reliable standards and getting good resolution. The *precision* refers to the error on a series of replicate determinations. *Reproducibility* is the ability to obtain the same result from day to day, place to place or even analyst to analyst. Thus precision is subject to systematic errors and reproducibility to random errors. Precise and accurate quantitative results are only obtained if due attention is paid to all phases of the analysis from first sampling to final result.

The four steps to be considered are:

1. sampling
2. chromatographic separation
3. detection
4. measurement and calibration

The success or otherwise in controlling these steps is often summed up in terms of a statistical analysis of the results, but for this the reader is referred to standard texts on statistics.

6.7.1 Sampling and Sample Preparation
In preparing any sample for analysis the usual procedures for a sound analytical technique should be followed. The sample should be homogeneous in the sense of being representative of the batch sample, and all samples should be prepared in the same manner. Samples must be completely soluble and the solvent used to dissolve the sample should, if possible, be the initial mobile phase. This minimizes changes in band shape and k' values that could reduce the precision and it also eliminates the possibility of the sample precipitating out on injection. If this is not possible the sample solvent must be miscible with the mobile phase and sample sizes should be restricted (< 25 μl). The sample solvent then should give a minimal detector signal and should be of high purity to avoid contamination problems. As mentioned earlier, the use of gradient elution may generate 'ghost peaks' and these should be checked for by running a blank gradient.

6.7.2 Chromatographic Separation
Decomposition of samples on the chromatograph is not usually a problem in

liquid chromatography but if the expected relationship between sample concentration and detector response is not observed then sample degradation should be suspected.

Potential errors are introduced by tailing peaks and by poor resolution. Tailing peaks result from poor columns or from incompatible mobile phases. If sufficient resolution is not achieved it is necessary to make assumptions about the unresolved portion of the recorder trace. Because of the symmetrical shape of the peaks generally obtained in liquid chromatography separations, extrapolation of the resolved portion of the recorder trace followed by peak height measurement or triangulation may often be adequate. However, a much higher level of precision and accuracy will be obtained using electronic integrators.

The new generation of computing integrators are able to cope with a wide range of operating conditions but they are not a substitute for a well-resolved chromatogram.

One of the main sources of error arises in the sample injection. A higher level of reproducibility can be achieved with a sample loop injector than with a syringe injector. With care, however, the syringe can be used to give satisfactory reproducibility and it has the advantage that the sample size can be varied without changing sample loops. The main causes of error in syringe injection arise from air bubbles in the syringe and from losses caused by a leaking septum or leaking syringe. The precision for peak area may be taken as 1–2 per cent with syringe injection and $< \pm 1\%$ for valve injection. The main reasons for this are:

1. the injected volume is less dependent on operator skill since it comes from a fixed volume sample loop.
2. the rate of injection is more reproducible so that sample introduction is constant.
3. the volume injected from an injection valve is usually larger so that the errors are less significant.
4. valves can be automated, again increasing precision.

Because the detectors in common use are concentration dependent, flow rate variations cause changes in both peak height and peak area measurements, as well as in peak shape. This increases the standard deviation of the measurement, though peak height is less dependent on flow rate than peak area.

If the characteristics of the column change, retention times and peak sizes will vary. Adsorbents and chemically bonded stationary phases are more stable than liquid stationary phases. Changes in the solvent composition will also affect retention times and peak heights, and solvent mixtures should be kept covered to reduce loss by evaporation of the more volatile components. If gradient elution is being used it is important to start the gradient at the same time for each analysis and to return the column to the initial conditions after each analysis. Isochratic analysis gives higher precision than gradient elution since it avoids the foregoing problems. Changes in column temperature will also lead to retention time and peak height changes, but the greatest errors caused by temperature changes occur when the detector temperature is not kept constant. The u.v. detector is not particularly susceptible to temperature changes but the refractometer detector is temperature sensitive and for accurate work must always be thermostatted (see Table 3.2).

Best quantitative results are obtained if the sample load is well below the column capacity so that retention times and plate heights are independent of sample size. Highest column efficiency and peak-height reproducibility occur at about 10^{-5}-10^{-6}g of solute per gram of packing for most types of packing [5]. With large sample sizes peak height measurements may not be linear.

6.7.3 Detection

The response obtained from a given detector will vary according to the nature of the solute molecule, and the chromatographer must calibrate the detector by running 'standards' at several levels to determine the response factor.

With a u.v. detector the response is related to both concentration and the molar extinction coefficient of the component at the wavelength of detection. With a differential refractometer the response is proportional to the concentration and the refractive index difference between the sample component and the mobile phase. The correct choice of the pertinent operating parameter, therefore, will not only increase the sensitivity of the detector but also provide greater precision of measurement in quantitation.

Detector sensitivity, baseline stability and linearity are important parameters in quantitative analysis and may affect the choice of detector, e.g. the sensitivity of the refractometer may not be high enough to detect trace components and specific detectors such as the u.v. detector may fail to detect components. The linear range of detectors may limit the accuracy obtainable unless the response is carefully calibrated. A large background from the mobile phase can reduce the linearity and detection limits of the detector.

Finally, a basic understanding of electronics is important in obtaining the best results from the detector. The effect of important specifications such as sensitivity, specificity, linearity, response time and dead band are often overlooked.

6.7.4 Measurement and Calibration

The relative merits of peak height and peak area measurements have been discussed in the literature [6–8]. In general, peak area measurements should be used where control of mobile phase composition is poor, for distorted peaks and when operator and instrument variations (other than flow rates) are large. Precise flow control is necessary if peak area measurements are used. Peak height measurement should be used if flow rate control is poor and for trace components. Peak height measurements are more accurate than peak area measurements for overlapping peaks.

Measurement of peak height and peak area. Peak height is measured from the baseline to the peak maximum. On a changing baseline a baseline must be interpolated from start to finish of the peak. Peak heights should not be used for distorted peaks or shoulders. Measurement of peak height is simple and rapid and can give a precision of 1–2% so that for many purposes it is sufficient.

Peak area may be measured in several ways:

1. Triangulation. Since chromatographic peaks approximate to triangles the area may be found by drawing tangents along the sides of the peaks to cut the baseline (Fig. 2.1). The peak height is then measured from the baseline to the point of intersection of the tangent and the width is given by their intersection with the baseline. Then $A = h.\frac{1}{2}w$. Alternatively the peak width is measured at half peak height $w\frac{1}{2}$ (Fig. 2.1) and $A = h. w\frac{1}{2}$.

   Neither method gives a true area but providing the peaks are not badly distorted the area is proportional to sample size. Both methods benefit from an increased chart speed to increase the peak width but in the first method this does increase the error due to a misplaced tangent.

2. Planimetry. The planimeter is a mechanical device which traces the perimeter of a peak and the area is measured digitally on a dial. The method is tedious and very dependent on operator skill, but it can be used with asymmetric peaks.

3. Cut and Weigh. In this method the peak is cut out and weighed. The precision depends on the weight constancy of the chart paper (and ink), but precision can be improved by photocopying the trace onto a higher grade of paper. This has the advantage of preserving the original trace. The method is superior to triangulation for irregularly shaped peaks.

4. Integrators. The use of integrators in quantitative analysis is widespread. These may vary from the simple ball-and-disc type mechanical integrator to highly sophisticated computing integrators.

   The ball and disc integrators give good accuracy independent of peak shape. Base-line drift and unresolved peaks can be corrected for, but this reduces the accuracy. There is no provision for peaks which go off scale.

   Digital integrators sense peaks and print-out areas in digital form. Sharp peaks are most easily handled. They are not able to assign baselines and detect baseline drifts and trends, so that wrong areas may be given to unresolved and tangent peaks. It is not necessary for recorder traces to be kept on-scale and since many of them also record retention times a recorder is not, in fact, essential. However, in order to set the correct parameters it is necessary to know parameters obtained from the chromatogram, such as the peak width. Generally, negative peaks cannot be handled (e.g. from a refractive index detector). The areas obtained are the 'raw' peak areas and corrections for detector response must be made manually. Computing integrators have both memory and computing capabilities so that they can be programmed to carry out certain functions such as the calculation of detector response factors and composition calculations. They are able to detect and correct for baseline changes and also to update the integrating parameters so that the accuracy is maintained as the eluting peaks become broader. They are also better equipped to handle poorly resolved peaks. A recent review of this subject is recommended to the interested reader [9].

To summarize:

(i) peak height measurement is preferred for manual data reduction because it is simple, fast and accurate. However, it is more susceptible to variations in separation conditions (except flow rate) than peak area measurements. It also requires less resolution of components and is best suited for analysis of multicomponent mixtures and for trace analysis.
(ii) the height times width at half-height method is the preferred manual method for area measurement for symmetrical peaks, but the ball-and-disc integrator, planimeter and cut and weigh methods, in decreasing order, are preferred for asymmetric peaks.
(iii) digital and computing integrators give the fastest and most accurate measurement of peak areas.

Calibration. Having obtained the raw data of peak heights or areas from the chromatograph or the integrator, this information must be converted into composition. Four methods are used: area normalization, external standard, internal standard and standard addition.

1. *Area normalization.* This technique is simple but in liquid chromatography it is of limited value, although it is often used in gas chromatography. It assumes that all components have been eluted and detected. The area of every peak is measured and this area is expressed as a percentage of the sum of the areas of all the peaks,

e.g.
$$\%X = \frac{A_X \times 100}{A_X + A_Y + A_Z}$$

In liquid chomatography this method only gives meaningful results if the detector response factors are calculated for each component. The response factor is calculated with respect to a reference compound (r) by

$$f_X = \frac{(A_r)(W_X)(f_r)}{(A_X)(W_r)},$$

where A_r and A_X are the areas of the reference and solute peaks respectively. W_r and W_X are the concentrations of the reference and solute compounds and f_r is the response factor assigned to the reference compound. With a u.v. detector f_r can be assigned by reference to its molar extinction coefficient but the simplest way is to assign it a value of unity.

The concentration of compound X is now given by

$$\%X = \frac{(A_X)(f_X) \times 100}{A_X f_X + A_Y f_Y + A_Z f_Z}$$

2. *External Standard.* A calibration plot is constructed by injecting a fixed volume of standard samples containing known weights (concentrations) of the compound of interest. A plot of peak height (area) versus weight (or concentration) for each compound is obtained from which the calibration factor S (peak size/weight or peak size/concentration) is thus obtained. (Fig. 6.7)

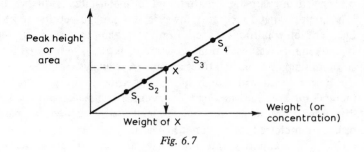

Fig. 6.7

For an unknown sample the amount of the component of interest X can be read off the graph against its area (A_X) or obtained from

$$\text{wt of X} = \frac{(\text{height})_X}{S}$$

The concentration of X in the original sample is then given by

$$\text{wt\% X} = \frac{\text{wt of X} \times \left[\dfrac{\text{total sample volume}}{\text{injected sample volume}}\right]}{\text{wt of original sample}}$$

The plots should be linear and pass through the origin. For non-linear plots the calibration factor cannot be used but the concentration of the unknown can be read off the plot. Non-linear plots require more calibration points than linear ones.

Peak height analysis can sometimes give non-linear plots where the peak area plot is linear. This is usually due to tailing peaks or column overload. In this case peak areas should be used.

A major source of error in this method is the injected sample volume Syringe injection is worse in this respect. With a valve injection overall precision better than 1 per cent is possible with the external standard method.

The calibration should be checked at regular intervals and the separation factor S updated to allow for changes in instrument parameters.

3. *Internal Standard.* This technique minimizes errors due to sample preparation, apparatus and technique. A known compound of fixed concentration is added to the unknown sample to give a separate peak in the chromatogram. From the peak areas of the sample and the standard the composition may be determined.

In GC the internal standard method is widely used to correct for injection volume errors. In LC using valve injectors this is not such a serious problem.

However, the technique is useful when lengthy sample preparation is involved (e.g. in derivatization) when recoveries may be variable. In this case the standard must, of course, be structurally similar and undergo the same reactions as the sample.

Calibration curves are obtained by chromatographing suitable volumes of calibration mixtures containing the compounds of interest with a constant concentration of the internal standard. Peak heights (areas) of the compounds of interest are determined and the ratio peak height of compound (A_s)/peak height of internal standard (A_{is}) is plotted against concentration of compound. (Fig. 6.8).

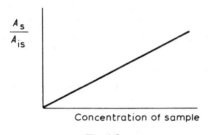

Fig. 6.8

The plot again should be linear and pass through the origin.

Response factors may be obtained by assigning a response factor of unity to the internal standard when

$$f_X = \frac{(A_X)(W_{is})}{(A_{is})(W_X)}$$

where A_X, W_X, A_{is}, W_{is} are the areas and weights of sample X and the internal standard respectively.

The sample composition may then be calculated from

$$\%X = \frac{(A_X)(f_X)(W_{is}) \times 100}{(A_{is})(W_X)}$$

With careful calibration precisions of 0.5% or better or obtainable.

The selection of the internal standard is most important and several requirements are placed on the standard:

(i)   it must be completely resolved from other peaks.
(ii)   it must elute near peaks of interest
(iii)   it should be added at a concentration which will give a similar peak height (area) to the compound of interest (taking into account different detector responses)
(iv)   it must not react with any other components
(v)   it must not be present in the original. sample
(vi)   should be of high purity and readily available

(vii) more than one internal standard may be necessary for multicomponent mixtures to give the highest precision

(viii) it must behave like the sample during any pretreatment

With complex mixtures it may not be possible to 'fit-in' a standard without overlapping another peak. In this case an external standard method should be used. Usually the standard is structurally related to the sample (e.g. an isomer or homologue), but if this is not possible it should have a similar solubility and also detector response.

4. *Standard Addition.* This method is only suitable when determining one or two components. The sample is first chromatographed then known amounts of the compounds of interest are added and the mixture re-chromatographed. The difference in peak heights (areas) in the two chromatograms then corresponds to the amount of sample added from which the amount of sample corresponding to the original peak height (area) can be calculated. By using a peak in the chromatogram whose composition is not to be measured as an internal standard, injection errors may be corrected for. In the absence of any such peak an internal standard can be added. This method is different from the three previous ones in that it tends to cancel out any matrix effects.

References
1. Majors, R. E. (1980) *J. Chromatog. Sci.*, **18**, 488.
2. Glajch, J. L., Kirkland J. J., Squire, K. M. and Minor J. M. (1980) *J. Chromatog.*, **199**, 57.
3. Aitzetmuller, K. (1975) *J. Chromatog.*, **125**, 43.
4. Sinsel, J. A., La Rue, B. M. and McGraw, L. D. (1975) *Anal. Chem.*, **47**, 1987.
5. Done, J. N. (1976) *J. Chromatog.* **125**, 43.
6. Ross, R. W. (1972) *J. Pharm. Sci.*, **61**, 1979.
7. Bakalyar, S. R. and Henry, R. A. (1976) *J. Chromatog.*, **126**, 327.
8. Kirkland, J. J. (1974) *Analyst*, **99**, 859.
9. Pierson, H. L. and Steible, D. J. (1977) in *Modern Practice of Gas Chromatography* (ed. R. L. Grob), Wiley-Interscience, NY, Chapter 8.

7

Preparative high performance liquid chromatography and trace analysis

7.1 Introduction

The characteristics of a chromatographic system can be represented by a tetrahedron [1] (Fig. 7.1). In analytical HPLC, resolution is the prime requisite, with speed of analysis another important variable. The scope of the system, i.e. its ability to separate mixtures of wide polarity range, is also important and can be varied by altering the mobile phase. Thus, the face of the tetrahedron joining speed, resolution, and scope is the most significant one for analytical HPLC but the capacity is sacrificed. The face of the tetrahedron joining capacity, scope, and resolution is more important for preparative HPLC, where speed and resolution are sacrificed to provide the maximum capacity.

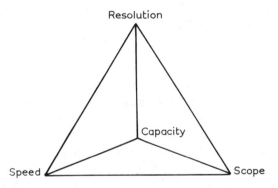

Fig. 7.1   Characteristics of a chromatographic system [1].

There are some differences of opinion as to what constitutes preparative HPLC. Some workers seek to distinguish between scale-up operations where analytical chromatographic conditions are used but with slightly larger sample loads, and large scale operations where the column is used under overloaded conditions. Both conditions will be described as preparative HPLC, but with reference to 'scale-up' or 'overloaded' conditions where appropriate.

There is a tendency for an analyst to use the technique with which he is most familiar, whatever the problems. It is salutory to remember that preparative HPLC is only one weapon in the analyst's arsenal. Depending on the nature of the solutes to be separated, it is sometimes better to use an alternative technique, e.g. distillation or fractional crystallization, rather than always to try preparative HPLC. Table 7.1 shows how the technique used will depend primarily on the amount of solute mixture to be tackled.

Table 7.1   *Comparison of separative procedures*

| | Sample size | | | |
	>50 g	>1 g	>1 mg	<1 mg
Distillation	✓	✓		
Crystallization	✓	✓	✓	
Counter-current Distribution	✓	✓		
Chromatography				
(a) Column	✓	✓	✓	
(b) Ion exchange	✓	✓	✓	
(c) Exclusion	✓	✓		
(d) Gas–liquid		?	✓	✓
(e) Thin layer			✓	✓
(f) Paper			✓	✓
(g) High performance liquid		?	✓	✓

7.2   Stages in Preparative HPLC
The steps involved in successful preparative HPLC are: (a) definition of the problem; (b) preparation of the sample for HPLC; (c) development of a separation on an analytical scale; (d) alteration of the variables to permit a large scale separation.

7.2.1   Single Component Mixture
(a) There are three basic situations where HPLC can be used on a preparative scale, and they are shown in Fig. 7.2. The easiest situation is where there is only one major constituent in the mixture, and it is the component that has to be obtained in good yield and high purity [Fig. 7.2(a)]. In the second example, the mixture has two major components [Fig. 7.2(b)], both of which need to be isolated. The mixture in the third example has several constituents, all of which are present in similar quantities [Fig. 7.2(c)]. To define the problem, it is necessary to recognize from the chromatogram that the mixture is like the first of the three situations in Fig. 7.2, i.e. it is a single component mixture whose major constituent is the required solute.

(b) The sample must be in a form suitable for HPLC, i.e. any impurities that could create problems should be removed (as in Chapters 4 and 5). To remove carboxylic acids from a mixture containing esters, it may be necessary to extract the acids from an ethereal solution of the solutes with aqueous sodium bicarbonate solution. A preliminary crystallization may be required to remove aromatic impurities from a reaction mixture so that an ultraviolet detector can be used to monitor the column eluent. Occasionally, a derivative has to be made that is suitable for one type of detector, usually to make the solute molecule absorb in the u.v. region.

(c) The next stage is to consider the analysis as a purely analytical problem to be solved, with the best resolution, in as short a time as possible. Normally, this

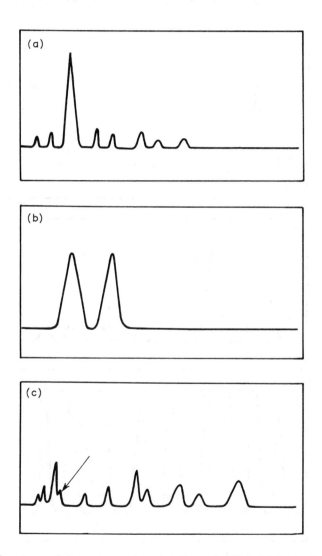

Fig. 7.2 Three types of chromatogram in preparative scale HPLC. (a) One major component mixture. (b) Two large component mixture. (c) Several component mixture.

analytical separation will be performed on a small microporous particle packing. It would be necessary to achieve a separation as in Fig. 7.3(a).

(d) To convert this separation into a preparative scale, it is essential to improve the resolution, but with a possible sacrifice in time. For the single component system [Fig. 7.3(b)], it is advisable to try to get as big a difference as possible in time between the preceding peaks and the major component, and between the major component and the succeeding peaks. When the major component is well separated from the other solutes, it is possible to introduce larger quantities of

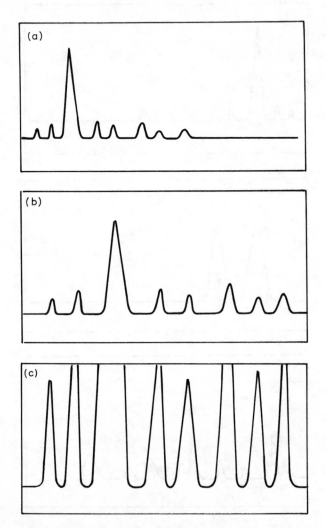

Fig. 7.3 Stages in scale-up of preparative HPLC. (a) Analytical separation. (b) Increased separation around the desired peak. (c) Overload conditions.

the sample on to the column. It is sometimes recommended that the solutes should have k' values > 5 so as to increase sample loadability.

7.2.2 Sample size and loadability

It has been found that sample loop injectors are better for preparative scale HPLC than syringe injectors, because larger volumes can be used and rubber from the septum is *not* introduced. In addition, problems with syringe injection of large volumes, which may require the flow to be stopped, are avoided. The volume of the sample injected on to the column can influence the efficiency markedly. Whilst it is usual to inject a fairly concentrated solution on to the column, there are occasions when the solubility of the solute requires a fairly dilute solution. In one trial, the analysis of a steroid (40 μg) on a 28.5 cm x 18 mm column showed that a twenty-fold increase of sample volume (0.1 cm^3 to 2.0 cm^3) led to a decrease in efficiency from 34.5 μm height equivalent to a theoretical plate, to 49.5 μm [2]. In any event, the sample volume should *not* exceed one-third of the mobile phase volume needed to elute the solute.

Table 7.2 shows that an increase in column diameter enables substantially more material to be analysed. More support provides greater surface area with

Table 7.2 *Typical sample sizes for preparative liquid–liquid and liquid–solid chromatography*

Column internal diameter (mm)	Relative internal cross-sectional area	Typical sample load (mg)		Injection volume	Solvent flow rates (cm^3 min^{-1})
		Easy separation	Difficult separation		
2.1	1.0	10–50	1	5–100 μl	0.3–6.0
6.2	8.0	100–500	10	0.5–5.0 cm^3	1.0–10.0
23.5	140.0	1000–5000	100	4–40 cm^3	5.0–90.0

which the solutes can interact, and since it does not increase the length of the column it does *not* increase the retention times. In the single major component system separation is relatively easy so that, even on a 2.1 mm i.d. column, 50 mg of sample could be separated when the column is overloaded. Some manufacturers have recommended the use of wider-bore columns which fit into an analytical instrument. Thus Whatman have shown that Partisil 10 can be used in columns up to 22 mm i.d. [3]. To give the same separation in the same time, a 1 cm^3/min flow rate for an analytical column (4.6 mm $\times$ 25 cm) must be changed to 4 cm^3/min flow rate on a 9.4 mm i.d. $\times$ 25 cm column and to 20 cm^3/min on a 22 mm i.d. $\times$ 25 cm column.

Where the resolution is large enough, columns can be used and overload samples can be analysed; e.g. on a 1 inch column samples of up to 5 grams can be injected. Some of the conditions for preparative scale HPLC are given in Table 7.2, and for exclusion chromatography in Table 7.3.

With increasing sample size, values of the height equivalent to a theoretical plate increase and the capacity ratios (k') decrease if other parameters are fixed. The performance of a preparative column with regard to loading, i.e. loadability, cannot be defined precisely but there are two definitions of maximum sample size which can be estimated quantitatively: (a) the linear capacity corresponds to the value of W_s/W_a which gives 10% decrease in the retention time of any solute compared with its retention time at the minimum detectable amount (where W_s is the weight of sample and W_a is the weight of adsorbent); or (b) the sample size where the h value is doubled compared with the h value determined with the minimum detectable sample size.

Using definition (a), acceptable sample sizes are much lower, and it is found that the maximum sample size for porous silica is about 2×10^{-4}g of sample

Table 7.3 *Typical sample sizes for preparative exclusion chromatography.*

Column outside diameter (inches)	Relative internal cross-sectional area	Typical column length (feet)	Typical sample load (g)		Typical injector volume (cm³)	Typical solvent flow rate (cm³ min⁻¹)
			easy separation	difficult separation		
3/8	1 ×	16	2	1	5–10	0.5–6.0
1.0	8 ×	16	18	9	45–90	3–30
2.4	60 ×	16	120	60	300–600	20–200

per gram of silica. The weight of stationary phase (and thus, sample size) can be increased by increasing either the column diameter or the column length. However, if the column is too short (less than 4 cm) the inlet and the outlet cause irregular movements in the solvent flow leading to band broadening. If the column is made too long, the packing at the inlet becomes overloaded, and there is a loss of resolution. If the internal diameter is increased, it becomes difficult to ensure that the solutes are spread evenly over the cross-section of the column at the inlet.

Endele *et al.* [4] claim that these measurements of loadability being proportional to the weight of the stationary-phase do not agree with the theory of the 'infinite diameter column' [5].

As important as the amount of packing in the column is the dead volume of the column since this indicates how much solvent is needed to equilibrate and to elute. In an analytical column where the dead volume is 3 cm³ and a solute has a k' value of 10, 30 cm³ of solvent has to be used. Equilibration can be achieved with 10-20 column volumes, i.e. 50 cm³. With a 22 mm × 50 cm preparative column where the dead volume is 130 cm³, 1300 cm³ of solvent would be needed to elute a solute with $k' = 10$ and 1.5-2.0 litres of solvent for equilibration.

7.2.3 Detector and Solvent Requirements

When the sample is overloaded as in Fig. 7.3(c) there may be difficulties with the detection system. Detectors are required which operate at high flow rates. In contrast to the analytical detectors, where special small bore tubing is used to provide minimum dead volume, preparative scale refractometers have large-bore tubing to carry the high flow rates. Alternatively, a stream splitter can take only a small aliquot of the column eluent into the detector, thus eliminating flow restrictions. The most popular detector for preparative scale HPLC is the refractometer, with the moving wire detectors a good second.

Having separated the components of the mixture, it is essential to have a good fraction collector so that resolution will not be lost by fraction mixing after the detector. One property of the solvent which was not stressed in Chapter 5 is the boiling point. In order to isolate the separated solutes from HPLC analyses, it is necessary to remove the solvent.

It is usual to remove the solvent by vacuum distillation. If the solvent has a very high boiling point, there is a danger that the solutes may decompose during the distillation. Low boiling solvents are therefore best for preparative HPLC.

A novel u.v. detector has been reported for preparative-scale HPLC [6] in which the solvent is allowed to flow over a supporting plate where it assumes the thickness characteristic of the flowing liquid film. The u.v. radiation falls on the film a little below the delivery tube from the column. The detector has been used for concentrations from 1 mg to 30 g and for flow rates from 1 cm³/min⁻¹ to 400 cm³/min⁻¹.

7.2.4 Two Major Component Mixture

The second type of problem is where there are two major components [Fig. 7.2(b)] which, on overload conditions, become one very broad band (Fig. 7.4). To obtain either solute in pure form, it is necessary to trap the front edge or the

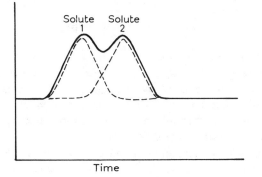

Fig. 7.4 Overload conditions for a two-component mixture.

rear edge of the combined band without collecting the material near the centre. It has been shown by GC–M.S. that the front edge is often pure component 1, whereas the middle of the band is contaminated with both solutes 1 and 2.

This situation can also be tackled by using the re-cycle technique. Partly resolved peaks can be returned to the column for further separation. The solutes pass through the column normally until the two partially resolved peaks are noted by the detector; then the 6-port valve 3 is switched to return these two peaks to the pump, for subsequent entry to the column for a second separation (Fig. 7.5). Nakanishi's separation of abscisic acid [7] shows such a separation. To resolve the diastereoisomers of abscisic acid, a 9 ft x 3/8 inch column was used, with a fully porous silica and a 100 mg sample. The first time the mixture was eluted from the column in 900 cm^3 of 1% propan-2-ol in hexane, the components were not resolved. The mixture was re-cycled five times, each time improving the resolution until complete resolution was achieved. By overloading the column with a 250 mg sample, satisfactory resolution was obtained after six passes through the column.

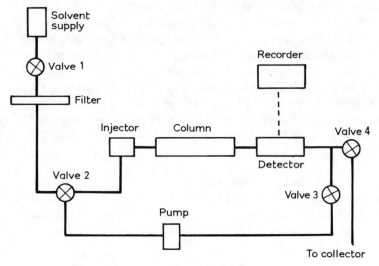

Fig. 7.5 Re-cycle technique.

7.2.5 *Several Component Mixture*
The third example [Fig. 7.2(c)] requires that preliminary fractions should be taken around the trailing edge of the large band to allow a cut to be obtained which has a higher concentration of the desired product. Treatment of the original mixture in this way should enrich the fraction until the problem reduces to the first example considered [Fig. 7.3(a)].

Another approach to preparative liquid chromatography is to use medium pressure systems (up to 300 lbf in^{-2} max.) with shatter-proof glass columns and Merck silica gel 60 (230–400 mesh). These columns can be dry packed with

100 g silica gel sealed with crimping seals and will separate 5 g of solute in 2-3 hours [8]. Solvent used does not need to be of analytical grade so long as it is distilled.

7.3 Infinite Diameter Columns

In the early days of liquid chromatography, it was believed that column efficiency decreased dramatically as the column internal diameter was increased, and large diameter columns were rarely used. It is now believed that this observation arose from inhomogeneous packing in those early wide-bore columns.

Knox and Parcher [5] described a phenomenon (the infinite diameter column) in which an injected band of solute never reaches the wall of the column before it has arrived at the column exit. They assumed that stream splitting is the important mechanism by which solute molecules placed in the centre of the column reach the walls, since diffusion of the solutes in liquids is very slow. Making this assumption, they described an expression relating column and packing dimensions to achieve the infinite diameter condition:

$$A = d_c^2 / d_p L \geqslant 2.4$$

where A is the Knox and Parcher coefficient, and d_c, d_p, and L are column internal diameter, particle diameter, and column length respectively. A regular increase in column efficiency with increasing values of A up to 45 has been observed [9].

De Stefano [10] has shown that increasing the diameter of the column from 2.1 to 4.76 mm i.d. produces a decrease in column efficiency. This is probably due to wall effects caused by increased particle segregation. However, columns with 7.94 mm and 10.9 mm internal diameters show improved efficiencies when the ratio of sample size to column cross-sectional area is kept constant. If the columns are used in an overloaded condition, the infinite diameter phenomenon does not apply.

7.4 Packing of a Preparative Column

Column packing can be a problem in wide-bore tubes, though not such a difficult one as in narrow-bore columns with small particles. One technique [11] uses a column (70 cm length x 18 mm i.d) with a piston which is equipped with a porous sintered stainless-steel disc capable of fitting tightly inside the column (Fig. 7.6). The piston is set to its lowest position and the injection port is removed. The packing material is poured into the column as a slurry in the solvent to be used for the analysis. The injector is replaced and packing is compressed by the piston forcing excess solvent through the 6-port valve. A pressure of 10 to 100 lbf in^{-2} is used for the packing, which can be completed in 10 minutes. This procedure gives a very homogeneous packing density, and eliminates dead volume near the injector. It is believed that this technique corresponds to the balanced density packing method used for narrow bore columns.

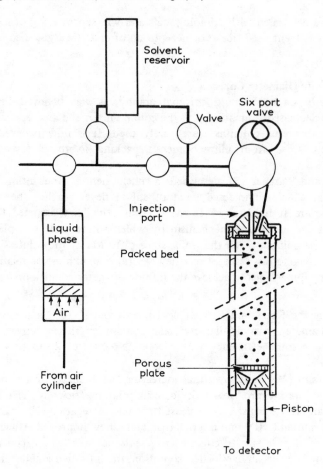

Fig. 7.6 Preparative scale column packing.

7.5 Summary of Preparative HPLC

It is clear that the requirements for preparative HPLC go considerably beyond those considered earlier for the analytical mode.

1. There must be an unlimited solvent supply.
2. It is best to use short but wide columns (length to diameter ratio of 20 or more).
3. Packing material should be porous rather than of pellicular type.
4. A reciprocating pump is best.
5. The pressure should be kept below 1000 lbf in^{-2};
6. Very high volume flow rates of solvent are required but linear velocity should be low.
7. Solvent should be volatile and have a low viscosity.

8. The sample should be dissolved in a relatively large volume of solvent.
9. Valve injectors are preferred to syringe injection.
10. The detector should be relatively insensitive and be capable of withstanding large solvent flows.
11. Overloading will occur where there is 1 mg of solute per cm^3 of packing; i.e. a 1-litre column should separate 1 gram of material.

7.6 Trace Analysis

HPLC can be used on an analytical scale (both qualitatively and quantitatively) and on a preparative scale. One very important area where HPLC can be employed is trace analysis. Indeed HPLC may supplant gas chromatography as the preferred mode for trace analysis, because it can tackle non-volatile, ionic, or labile solutes which cannot be analysed readily by gas chromatography.

To obtain reproducible and accurate trace analysis by HPLC, it is necessary to optimize the chromatographic sampling method, the calibration technique, the column resolution, and the detection.

7.6.1 Sampling

The sample should be injected on to the column in as large a volume as possible, and as a sharp band to make detection easy. Although the volume of the sample may be large, it should not be more than one-third of the volume of mobile phase needed to elute the solute molecules. A broad diffuse band formed by poor injection technique will give decreased concentration at the detector and thus reduced peak heights. At the other extreme, too large a weight of solute will overload the front of the column, producing poor column efficiency and resolution. Using analytical columns with i.d between 2 and 3 mm, samples up to 300 μl can be injected without changing the capacity factor or causing a loss in column performance. When the column is operated at high pressures (2000–5000 lbf in^{-2}), it may prove difficult to inject large samples by syringe unless the flow of the mobile phase is stopped. Alternatively, sampling valves (up to 5 cm^3) can be used for pressures up to 7000 lbf in^{-2}.

If the trace component is present at very low concentration, some form of pre-concentration may prove useful. However, care must be taken to ensure that the concentration of the major constituent does not become so large that it overloads the column, or that the trace component is altered in any way. Other techniques have been developed for situations where the solute is so insoluble that it cannot be concentrated by solvent evaporation, e.g. coupled columns and back flushing.

The passage through a short pre-column may provide only partial resolution but the analysis is accomplished in a short time (Fig. 7.7).

Only that portion of the mixture containing the partly resolved trace component is switched to the main analytical column by a low volume valve. Better resolution of this small portion of the sample is obtained in the longer analytical column. These separations usually require a constant volume pump.

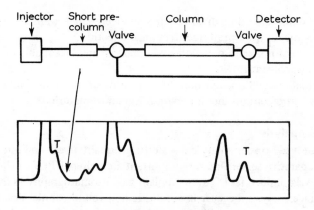

Fig. 7.7 Coupled column technique.

This type of procedure involves a considerable effort and is mostly applied on a routine basis, where the large number of trace analyses justifies the effort expended.

As explained earlier, natural product mixtures often contain components of widely differing polarity (see Fig. 5.2). The long elution times with solvents of increasing eluent strength, as well as the equilibration time to return the column to its original condition, make trace analysis tedious and time-consuming. Back-flushing is one procedure that can be applied to minimize these difficulties. The sample is injected on to the column and, after the trace component has eluted, the direction of flow of the mobile phase is reversed so that the more strongly retained components of the mixture will be flushed from the column to waste (Fig. 7.8).

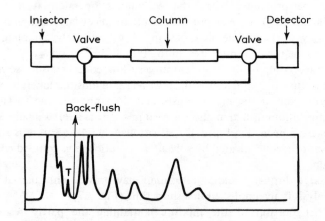

Fig. 7.8 Back-flushing technique. Back flushing at the point indicated allows the peaks that would be eluted later to be back-flushed from the column.

7.6.2 Calibration

In order to develop calibration curves for trace analysis, it is essential that the accuracy of measurement is good, but the precision need not be so high. Indeed, chromatographic reproducibilities of about 10% are adequate. Peak height is measured rather than area because it is less influenced by overlapping peaks. It can be shown that, where $R_S = 1$, the concentration of two bands can change by a factor of 1000, with only a 3% error in the measurement [12]. Straight line calibration curves are usually found for trace analysis.

7.6.3 Resolution

Peak height measurement is easier and more accurate if the trace component is eluted before the major component. When the trace solute has a longer retention time, it will be eluted in the tail of the major component, or even be masked completely. Chromatographic conditions should be altered to permit the former situation to occur if possible.

The peak height of the trace component should be made as large as possible by improving column resolution.

The resolution can be optimized by varying the separation factor. Peak height measurement can be made easier if the band centre of one solute is displaced relative to another, i.e. if the separation factor is changed by altering the mobile or the stationary phase.

Increasing the number of theoretical plates of the column can produce sharper peaks which are more easily measured. For trace analysis it is better to use lower mobile phase velocities, and satisfactory results are obtained with 0.4 and 0.1 cm s^{-1} for 10 and 40 μm respectively.

The sensitivity of trace analysis is affected by the capacity factor k', which must be adjusted to provide maximum peak height by altering the mobile phase. Increasing k' will provide improved resolution and accuracy, but unfortunately the peak height is usually decreased with concomitant lower accuracy in peak height measurement. In trace analysis, desired k' values are normally smaller than the range 2–5, which is accepted for the best analytical HPLC. If possible, the mobile phase should be altered to provide the k' value in the range 0.5–1.5 where the sensitivity is best.

7.6.4 Detectors

Ideally, it would be preferable to have a selective detector which responded only to the trace component and was unaffected by the major peak. The most satisfactory devices for trace analysis are: ultraviolet spectrophotometers, spectrofluorimeters, colour/reaction and polarographic detectors, of which only the latter cannot be used with gradient elution. Their approximate sensitivity for suitable trace solutes is 5×10^{-7} g l^{-1} for the u.v., 1×10^{-7} g l^{-1} for colour/reaction detectors, and $<1 \times 10^{-7}$ g l^{-1} for polarographic detectors.

7.6.5 Summary of Trace Analysis

In summary, trace analysis is best performed:

1. with a separation factor of 1.2 to provide easy resolution;
2. with a capacity factor between 0.5 and 1.5;
3. with mobile phase velocity less than 1 cm s^{-1} (between 0.1 and 0.4 cm s^{-1});
4. with large sample volumes;
5. with an efficient column having between 2000 and 4000 theoretical plates;
6. with a pulseless pump to improve signal-to-noise ratio;
7. with a selective detector to monitor the particular solute at a large signal-to-noise ratio.

References

1. Scott, R. P. W. and Kucera, P. (1974) *J. Chromatog. Sci.*, **12**, 473.
2. Godbille, E. and Devaux, P. (1976) *J. Chromatog.*, **122**, 317.
3. Rabel, F. M. (1980) *International Laboratory*, Nov., p. 91.
4. Endele, R., Halasz, I. and Unger, K. (1974) *J. Chromatog.*, **44**, 377.
5. Knox, J. H. and Parcher, J. F. (1969) *Anal. Chem.*, **41**, 1599.
6. Miller, J. M. and Strutz, R. (1980) *International Laboratory*, March, p. 87.
7. Koreeda, M., Weiss, G. and Nakanishi, K. (1973) *J. Amer. Chem. Soc.*, **95**, 239.
8. Bundle, D. R., Iverson, I. and Josephs, S. (1980) *International Laboratory*, Nov., p. 27.
9. Vermont, J., Deleuil, M., de Vries, A. J. and Guillemin, L. L. (1975) *Anal. Chem.*, **48**, 1329.
10. De Stefano, J. J. and Beachall, H. C. (1972) *J. Chromatog. Sci.*, **10**, 654.
11. Godbille, E. and Devaux, P. (1974) *J. Chromatog. Sci.*, **12**, 564.
12. Kirkland, J. J. (1974) *Analyst*, **99**, 859.

8

Applications of high performance liquid chromatography

In this chapter on applications of HPLC, the division into pharmaceuticals, biochemicals, food chemicals and industrial chemicals is somewhat arbitrary, with the result that there are examples where the chemical which has been analysed could be placed in two or three sections, e.g. vitamins could be included under biochemicals, pharmaceuticals or food chemicals. The industrial chemicals section includes the products of both the heavy chemical industry and the fine chemical industry.

The applications described in this chapter are intended to enable a beginner in HPLC to find similar compounds to those which he wishes to analyse and in conjunction with the understanding of the technique gained from the earlier chapters, to decide which column, which solvent and which detector he will use. No attempt has been made to indicate for each separation whether it is reverse phase, ion exchange etc. It is hoped that the reader will now be able to recognize this for himself. Readers requiring a more extensive treatment of the applications of HPLC are referred to the bibliography at the end of this introduction.

Each new technique goes through a phase when every researcher is keen to show pictorially how good his results are. Thus, during the 1930s, the journals allowed authors to include a graph of their ultraviolet spectrum. HPLC is just about to leave this stage in its development, so in this chapter, it is not considered necessary to show the actual chromatogram reported; instead retention times are given so that readers can get some idea of the relative elution order of the solutes under the specified conditions.

The data are presented in a standardized format which, it is hoped, researchers will follow in their own reports. It is surprising to find that temperature and pressure are not mentioned in many publications.

The *column* entry consists of the name of the packing material and its particle size, the column length in metres and its internal diameter in mm (where all these variables have been reported).

The *column temperature* is reported in degrees Centigrade. Where the temperature is left blank, none has been reported, but it probably means ambient temperature.

The *mobile phase* heading is followed by the solvent mixture with the relative proportions of the solvents quoted, usually volume for volume.

The *flow rate* is given, in $cm^3 min^{-1}$, though many authors do not report this value.

The *pressure* heading is followed by the value quoted by the researcher together with that value calculated in pound-force per square inch. It would seem appropriate to standardize on $lbf in^{-2}$.

The *detector* heading is followed by an indication of the type of detector used i.e. u.v. 254 nm shows that ultra violet radiation at 254 nm wavelength was utilised. Where the u.v. wavelength is not quoted, it is probably at 254 nm.

The *sample* heading is followed by the compound(s) and their retention times.

Finally the *reference* heading is followed by the authors and the journal, except where the separation has been reported in applications sheets which have kindly been provided by Merck, A. G., Du Pont Instruments Co., Varian Instruments Co., Waters Associates, Jobling (Corning) Laboratory Division and Applied Research Laboratories.

It is debatable whether theoretical plates should be quoted in a standardized format but it was considered unwise to calculate this value from an author's published trace.

8.1   Pharmaceuticals

8.1.1   Alkaloids

(a) COLUMN   Microporous silica gel (10 μm) (0.25 m x 2.2 mm i.d.)
    COLUMN TEMP
    MOBILE PHASE isopropyl ether—acetonitrile—methanol
                 (69.5 : 30.0 : 0.5)
    FLOW RATE   0.33 $cm^3 min^{-1}$
    PRESSURE
    DETECTOR   fluorescence
    SAMPLE   reserpine   5.6 min
                    ergotaminine 7.5 min

    REFERENCE   R. J. Perchalski, J. D. Winefordner and B. J. Wilder
                    (1975) *Anal. Chem.* **47**, 1993.

(b) COLUMN   Merckosorb SI 60 (5 μm) (0.20 m x 2.0 mm i.d.)
    COLUMN TEMP
    MOBILE PHASE chloroform—methanol (85 : 15)
    FLOW RATE   1 $cm^3 min^{-1}$
    PRESSURE   197 bar (2860 $lbf in^{-2}$)

DETECTOR   u.v. 254 nm
SAMPLE   emetine hydrochloride   60 s
              cephaeline hydrochloride   80 s

REFERENCE   Merck, A. G., Darmstadt, W. Germany,

(c) COLUMN   Merck RP-8 7 μm (25 cm $\times$ 2.0 mm)
    COLUMN TEMP
    MOBILE PHASE methanol–water–formic acid (166 : 34 : 1) buffered with
                  triethylamine, pH 8.5
    FLOW RATE   1.0 cm^3 min^{-1}
    PRESSURE   104 atm (1670 lbf in^{-2})
    DETECTOR   u.v. 330 nm
    SAMPLE   Alkaloids   harmol   3.4 min
                              harmine   4.2 min
                              harmaline 4.8 min

    REFERENCE   F. Sasse, J. Hammer and J. Berlin (1980) *J. Chromatog.*
                  194, 234.

8.1.2   Antibiotics
(a) COLUMN   Zipax (37 μm) + 1% cation exchange resin (SCX)
                  (1.5 m x 2 mm i.d.)
    COLUMN TEMP
    MOBILE PHASE 0.01M-EDTA + 0.01M-KH$_2$PO$_4$ at pH 7.0
    FLOW RATE   1 cm^3 min^{-1}
    PRESSURE
    DETECTOR   u.v.254 nm
    SAMPLE   Oxytetracycline   11 min

    REFERENCE   Varian Associates

(b) COLUMN   Pellionex SCX 30–40 μm (1.5 m x 2.2 mm i.d.)
    COLUMN TEMP   45°C
    MOBILE PHASE 0.1M-NH$_4$NaHPO$_4$ (pH 8.2)
    FLOW RATE   0.8 cm^3 min^{-1}
    PRESSURE   50 atm (750 lbf in^{-2})
    DETECTOR   u.v. 280 nm
    SAMPLE   Oxytetracycline HCl   2.4 min
                  Tetracycline HCl   3.8 min
                  Deoxycycline HCl   4.6 min

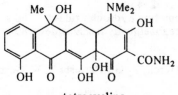

tetracycline

REFERENCE   Varian Associates

(c) COLUMN   Micropak SI 10 (0.15 m x 2 mm i.d.)
    COLUMN TEMP
    MOBILE PHASE   isopropanol–0.5M-acetate buffer at pH 4.5 (90 : 10, v/v)
    FLOW RATE $0.4 \text{ cm}^3 \text{ min}^{-1}$
    PRESSURE
    DETECTOR   u.v.
    SAMPLE   Daunomycin II   1.5 min
                   Daunomycin I   2.5 min

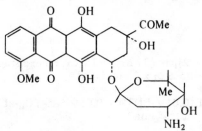

REFERENCE   Varian Associates

(d) COLUMN   Aminex A 28 (1.0 m x 2 mm i.d.)
    COLUMN TEMP $40°C$
    MOBILE PHASE   water
    FLOW RATE $0.2 \text{ cm}^3 \text{ min}^{-1}$
    PRESSURE
    DETECTOR
    SAMPLE   Streptomycins:
                   Deoxykanamycin   5.0 min
                   Kanamycin A   6.2 min
                   Kanamycin B   14.6 min

	R_1	R_2	R_3
A	OH	OH	OH
B	NH_2	OH	OH
DKB	NH_2	H	H

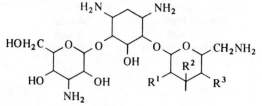

REFERENCE   Varian Associates

(e) COLUMN   Micropak CH 10 μm (0.25 m x 2.1 mm i.d.)
    COLUMN TEMP ambient
    MOBILE PHASE 30% methanol + 70% (1% acetic acid in water)
    FLOW RATE   2 cm³ min⁻¹
    PRESSURE   380 atm (5700 lbf in⁻²)
    DETECTOR   u.v. 268 nm
    SAMPLE   Penicillin V acid   3.6 min

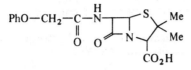

    REFERENCE   Varian Associates

(f) COLUMN   Micropak CH 10 μm (0.25 m x 2.1 mm i.d.)
    COLUMN TEMP 50°C
    MOBILE PHASE 10% methanol in water with 0.01M-phosphoric acid
    FLOW RATE   1 cm³ min⁻¹
    PRESSURE   70 atm (1050 lbf in⁻²)
    DETECTOR   u.v. 260 nm
    SAMPLE   Cephalexin   2.6 min

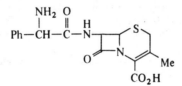

    REFERENCE   Varian Associates

(g) COLUMN   Lichrosorb RP 8 10 μm (25 cm × 3.2 mm i.d.)
    COLUMN TEMP
    MOBILE PHASE Phosphate buffer (pH 7.0)–methanol (6 : 4)
    FLOW RATE   1.5 cm³ min⁻¹
    PRESSURE   120 bar
    DETECTOR   u.v. 220 nm
    SAMPLE   Penicillin V   3.3 min

    REFERENCE   F. Nachtmann (1979) *Chromatographia* **12**, 380

(h) COLUMN   Lichrosorb RP 2 5 μm
    COLUMN TEMP
    MOBILE PHASE acetonitrile–water–0.1 M-phosphoric acid (20 : 70 : 10)
    FLOW RATE   0.8–1.0 cm³ min⁻¹
    PRESSURE
    DETECTOR   fluorescence
    SAMPLE   Adriamycin (doxorubicin)   6.0 min
                  Adriamycinol metabolite   3.5 min

    REFERENCE   R. N. Pierce, P. I. Jatlow (1979) *J. Chromatog.* **164**, 471

(i) COLUMN   Bondapak C_{18} (30.4 cm × 4.0 mm i.d.)
    COLUMN TEMP
    MOBILE PHASE methanol–0.01 M-tetrabutylammonium bromide (4 : 7)
    FLOW RATE   3.0 cm^3 min^{-1}
    PRESSURE   233 atm (3500 lbf in^{-2})
    DETECTOR   u.v. 254 nm
    SAMPLE   Semisynthetic penicillins
                  Carbenicillin   5.5 min
                  Diastereoisomers 6.5 min

    REFERENCE   K. Yamaoka, S. Narita, T. Nakagawa and T. Uno (1979)
                 J. Chromatog. **168**, 187.

(j) COLUMN μBondapak C_{18} (30 cm × 3.9 mm i.d.)
    COLUMN TEMP 60°C
    MOBILE PHASE 21% acetonitrile in aqueous KH_2PO_4 (pH 3.0)
    FLOW RATE   5.0 cm^3 min^{-1}
    PRESSURE   26.6 atm (400 lbf in^{-2})
    DETECTOR
    SAMPLE

Adriamycin	6.3 min
Adriamycinone	6.9 min
Dihydrodaunamycin	7.8 min
Dihydrodaunomycinone	9.3 min
Daunamycin	10.3 min
Daunomycinone	12.0 min

    REFERENCE   A. Alemanni and M. Riedmann (1979) *Chromatographia*
                 12, 396.

8.1.3   Drugs

(a) COLUMN   Sil-X-II (0.5 m × 2.6 mm i.d.)
    COLUMN TEMP
    MOBILE PHASE 1% methanol in chloroform
    FLOW RATE   0.8 cm^3 min^{-1}
    PRESSURE
    DETECTOR   u.v. 254 nm
    SAMPLE   Valium   3.0 min
                  Oxazepam 3.5 min
                  Librium   4.5 min
                  Flurazepam 7.5 min

    REFERENCE   D. H. Rogers (1974) *J. Chromatog. Sci.* **12**, 742.

(b) COLUMN   Corasil II 37–50 μm (0.5 m × 2.3 mm i.d.)
    COLUMN TEMP
    MOBILE PHASE 0.22% cyclohexylamine in cyclohexane
    FLOW RATE   1.45 cm^3 min^{-1}

PRESSURE
DETECTOR
SAMPLE   Phencyclidine   0.20
Methadone   0.23
Cocaine   1.00 (5.91 min)
Tetrahydrocannabinol   1.20
Valium   1.33
Methamphetamine   2.66

REFERENCE   M. L. Chan, C. Whetsell and J. D. McChesney (1974)
J. Chromatog. Sci. **12**, 512.

(c) COLUMN   SAX strong anion exchange (1.0 m)
COLUMN TEMP
MOBILE PHASE distilled water at pH 6 (phosphate buffer)
FLOW RATE   1 cm^3 min^{-1}
PRESSURE   1200 lbf in^{-2}
DETECTOR   u.v.
SAMPLE   Lysergic acid diethylamide (LSD)   4.5 min

REFERENCE   Du Pont Co. Ltd. Instrument Products Division.

(d) COLUMN   Magnum 9-ODS Partisil 10-ODS (25 cm × 9.4 mm i.d.)
COLUMN TEMP
MOBILE PHASE water–methanol–acetic acid (59 : 40 : 1) + 0.04 M-
methanesulphonic acid at pH 3.5
FLOW RATE   5.0 cm^3 min^{-1}
PRESSURE
DETECTOR   u.v. 254 nm
SAMPLE   LSD   17.3 min
iso-LSD   31.0 min

REFERENCE   I. S. Lurie and J. M. Weber (1978) *J. Liq. Chrom.* **1**, 587.

(e) COLUMN   Corasil II (0.5 m × 2.3 mm i.d.)
COLUMN TEMP
MOBILE PHASE 0.22% cyclohexylamine + 1.5% methanol in cyclohexane
FLOW RATE   1.45 cm^3 min^{-1}
PRESSURE
DETECTOR
SAMPLE   3,4-Methylenedioxyamphetamine   1.00 (4.14 min)
Secobarbital   1.48
Heroin   2.14
N, N-Dimethyltryptamine   2.52
LSD   3.81
Mescaline   6.66

REFERENCE   M. L. Chan, C. Whetsell and J. D. McChesney (1974)
J. Chromatog. Sci. **12**, 512

(f) COLUMN   Micropak SI-5 (0.25 m x 2 mm i.d.)
     COLUMN TEMP
     MOBILE PHASE hexane–dichloromethane–isopropanol–acetic acid
                        (97 : 2.2 : 0.6 : 0.2)
     FLOW RATE   1 cm^3 min^{-1}
     PRESSURE   3000 lbf in^{-2}
     DETECTOR   u.v. 254 nm
     SAMPLE   Secobarbital   13.9 min
                            Sandoptal   15.6 min
                            Amobarbital   19.0 min
                            Butabarbital   20.6 min
                            Diphenylhydantoin   26.1 min
                            Phenobarbital   30.4 min

     REFERENCE   Varian Associates

(g) COLUMN   Magnum 9 ODS Partisil 10-ODS (25 cm x 9.4 mm i.d.)
     COLUMN TEMP
     MOBILE PHASE water–methanol–acetic acid (59 : 40 : 1) + 0.05 M-
                        methanesulphonic acid at pH 3.5
     FLOW RATE   5 cm^3 min^{-1}
     PRESSURE
     DETECTOR   u.v. 254 nm
     SAMPLE   Amobarbital   8.0 min
                            Secobarbital   9.5 min

     REFERENCE   I. S. Lurie and J. M. Weber (1978) *J. Liq. Chromatog.* **1**,
                        587.

(h) COLUMN   H. C. Pellosil (Reeve Angel) (0.5 m x 3.0 mm i.d.)
     COLUMN TEMP
     MOBILE PHASE 10% isopropanol in dichloromethane
     FLOW RATE   3.0 cm^3 min^{-1}
     PRESSURE
     DETECTOR   u.v. 254 nm
     SAMPLE   Chlorpromazine   1.3 min
                            Metadopramide   6.0 min

chlorpromazine   metadoprimide

     REFERENCE   Applied Research Laboratories Ltd.

(i) COLUMN   Lichrosorb Si 60 5 μm (15 cm × 0.46 cm)
    COLUMN TEMP
    MOBILE PHASE 0.06 M-NaBr in ethanol
    FLOW RATE   1.2 cm^3 min^{-1}
    PRESSURE
    DETECTOR   u.v.
    SAMPLE   Antidepressant basic drugs as ion pairs

Desmethylclomipramine	2.5 min
Trimipramine	2.8 min
Clomipramine	3.5 min
Imipramine	4.0 min

    REFERENCE   J. E. Greving, H. Bouman, J. H. G. Jinkman, H. G. M. Westenberg and R. A. de Zeeuw (1979) *J. Chromatog.* **186**, 683.

(j) COLUMN   Spherisorb 10% ODS (25 cm)
    COLUMN TEMP
    MOBILE PHASE 25% methanol in water
    FLOW RATE   1 cm^3 min^{-1}
    PRESSURE
    DETECTOR   u.v. 326 nm
    SAMPLE   Misonidazole   2.95 min

    REFERENCE   T. R. Marten and R. J. Ruane (1980) *Chromatographia* **13**, 137.

(k) COLUMN μBondapak C$_{18}$ (30 cm × 3.8 mm i.d.)
    COLUMN TEMP
    MOBILE PHASE methanol–water (60 : 40)
    FLOW RATE   2 cm^3 min^{-1}
    PRESSURE
    DETECTOR   u.v. 254 nm
    SAMPLE   8-Methoxypsoralene (methoxsalen)   3.5 min

    REFERENCE   A. H. Hikal, A. R. M. Morad and S. El-Houfy (1980) *Chromatographia* **13**, 105.

(l) COLUMN   Permaphase ETH
    COLUMN TEMP 50°C
    MOBILE PHASE 60% n-hexane–40% isopropanol
    FLOW RATE   1 cm^3 min^{-1}
    PRESSURE   300 lbf in^{-2}
    DETECTOR   u.v.
    SAMPLE

Sulphapyridine	2.5 min
Sulphamethazone	3.5 min
Sulphanilamide	6.5 min

    REFERENCE   Du Pont Co. Ltd. Investment Products Division.

(m) COLUMN   TEAE-Cellulose (1 m × 2.1 mm i.d.)
     COLUMN TEMP 80°C
     MOBILE PHASE 0.1 M-Tris(hydroxymethyl)aminomethane (THAM) at
                 pH 8.00
     FLOW RATE   0.2 cm^3 min^{-1}
     PRESSURE   400 lbf in^{-2}
     DETECTOR
     SAMPLE   Sulphanilamides

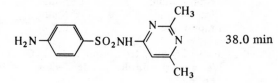
38.0 min

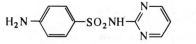
39.2 min

H_2N—⟨ ⟩—$SO_2NHCOCH_3$ 46.0 min

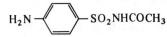
47.3 min

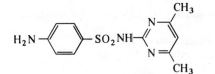
81.3 min

REFERENCE   W. Morozowich (1974), *J. Chromatog. Sci.* **12**, 453.

(n) COLUMN   μBondapak C$_{18}$ (30 cm × 4 mm i.d.)
     COLUMN TEMP
     MOBILE PHASE water–methanol–acetic acid (85 : 15 : 0.5) + 0.1% tetra-
                 butylammonium hydroxide
     FLOW RATE   1 cm^3 min^{-1}
     PRESSURE
     DETECTOR   u.v.

SAMPLE	Sulphanilamide	4.2 min
	Sulphanilic acid	7.6 min
	Sulphamethazine	16.0 min

REFERENCE R. C. Doerr, N. Parris and O. W. Parks (1980) *J. Chromatog.* **196**, 498.

8.1.4 Steroids

(a) COLUMN   Zorbax SIL (0.25 m x 2.1 mm i.d.)

COLUMN TEMP

MOBILE PHASE 0.1% methanol in hexane

FLOW RATE

PRESSURE   1500 lbf in^{-2}

DETECTOR   Refractometer

SAMPLE	Cholestane	0.50 min
	5α-Cholestan-3-one	0.75 min
	Cholesterol	3.0 min

REFERENCE   Du Pont Co. Ltd., Instrument Products Division.

(b) COLUMN   *N*-β(aminoethyl)-γ-aminopropylsilane on silica (Micropak NH$_2$) (0.15 cm)

COLUMN TEMP

MOBILE PHASE solvent A: 0.2% isopropanol in hexane

                  solvent B: 50% isopropanol in methylene chloride

FLOW RATE

PRESSURE

DETECTOR

SAMPLE	Progesterone	1.20 min
	Androstendione	1.50 min
	Δ^4-Pregnen-20β-ol-one	2.05 min
	17α-Hydroxyprogesterone	2.50 min
	Adrenosterone	2.80 min
	Cortisone acetate	3.60 min
	Cortisone	5.50 min
	Hydrocortisone	6.60 min

REFERENCE   R. E. Majors and M. J. Hopper (1974) *J. Chromatog. Sci.* **12**, 767.

(c) COLUMN   Merckosorb SI 60 (5 μm) (0.20 m x 2.0 mm i.d.)

COLUMN TEMP

MOBILE PHASE chloroform + 1% methanol

FLOW RATE   1.1 cm^3 min^{-1}

PRESSURE   197 bar (2860 lbf in^{-2}

DETECTOR   u.v. 254 nm

SAMPLE	Dexamethasone acetate	1.5 min
	Hydrocortisone acetate	2.1 min
	Epoxycortisone	3.2 min
	Cortisone	4.0 min
	Prednisone	5.3 min
	Prednisilone	16.3 min

REFERENCE   Merck A. G., Darmstadt.

(d) COLUMN   Poragel PN (0.9 m x 0.305 in i.d.)
      COLUMN TEMP
      MOBILE PHASE methanol—water (7 : 3)
      FLOW RATE
      PRESSURE
      DETECTOR

SAMPLE	Ecdysone	12 min
	Cyasterone	17 min
	Ponasterone	22 min

REFERENCE   Waters Associates

(e) COLUMN   Permaphase ETH (1 m x 4.0 mm i.d.)
      COLUMN TEMP
      MOBILE PHASE 60% (v/v) water—ethanol
      FLOW RATE   2.5 cm^3 min^{-1}
      PRESSURE   700 lbf in^{-2}
      DETECTOR   u.v.

SAMPLE	Norgestrel	5.0 min
	Ethinyl oestradiol	10.5 min

REFERENCE   Jobling (Corning) Laboratory Division.

(f) COLUMN   Varian SI 10 (25 cm × 2 mm i.d.)
      COLUMN TEMP
      MOBILE PHASE isopropanol–cyclohexane (1 : 9)
      FLOW RATE   1 cm^3 min^{-1}
      PRESSURE   50 kgf cm^{-2}
      DETECTOR   u.v. 240 nm
      SAMPLE   Hydrocortisone acetate   4.5 min

REFERENCE   A. R. Lea, J. M. Kennedy and G. K. C. Low (1980)
                J. Chromatog. **198**, 41.

(g) COLUMN   ODS Hypersil (15 cm × 4.6 mm i.d.)
    COLUMN TEMP   45°C
    MOBILE PHASE 40 to 100% methanol in water
    FLOW RATE   1 cm³ min⁻¹
    PRESSURE
    DETECTOR   240 nm
    SAMPLE   18, 21-Dihydroxypregn-4-ene-3, 11, 20-trione   11 min
                  Aldosterone   13 min
                  18, 21-Dihydroxypregn-4-ene-3, 20-dione 20,
                      18-hemiacetal   27 min
                  20 other steroids given

    REFERENCE   M. J. O'Hare, E. C. Nice and M. Capp (1980)
                  J. Chromatog. **198**, 22.

8.1.5   Analgesics
(a) COLUMN   Perisorb KAT (30–40 μm) (0.50 m x 2.0 mm i.d.)
    COLUMN TEMP
    MOBILE PHASE aq. buffer soln. (pH 8.5)
    FLOW RATE   0.6 cm³ h⁻¹
    PRESSURE   19 bar (275 lbf in⁻²)
    DETECTOR   u.v. 254 nm
    SAMPLE   Acetylsalicylic acid   1.1 min
                  Phenacetin   2.8 min
                  Caffeine   5.0 min

    REFERENCE   Merck A. G., Darmstadt,

(b) COLUMN   Zipax SAX anion exchange (0.90 m x 4.0 mm i.d.)
    COLUMN TEMP
    MOBILE PHASE 0.01M-borate buffer (pH 9.2) + sodium nitrate
    FLOW RATE   2.7 cm³ min⁻¹
    PRESSURE   500 lbf in⁻²
    DETECTOR   u.v.
    SAMPLE   Caffeine   1.83 min
                  Phenacetin   2.60 min
                  Aspirin   4.40 min

    REFERENCE   Jobling (Corning) Laboratory Division.

(c) COLUMN   Lichrosorb RP-8 10 μm (25 cm × 4.6 mm i.d.)
     COLUMN TEMP
     MOBILE PHASE 58% 0.01 M-sodium acetate–22% 1% acetic acid–20%
                   methanol
     FLOW RATE   2 cm^3 min^{-1}
     PRESSURE   120 atm (1804 lbf in^{-2})
     DETECTOR   u.v.
     SAMPLE   Theobromine   2.1 min
                    Theophylline   2.8 min
                    8-Chlorotheophylline   3.8 min
                    Caffeine   4.9 min

     REFERENCE   P. A. Reece and I. Cozamanis (1979) Spectra Physics
                    Review Vol. 5 no 2.

(d) COLUMN   Sil-X-1 ODS (25 cm × 2.6 mm i.d.)
     COLUMN TEMP   40°C
     MOBILE PHASE acetonitrile–0.2 M-phosphate buffer, pH 4.0 (1 : 9)
     FLOW RATE   1.0 cm^3 min^{-1}
     PRESSURE
     DETECTOR   u.v. 205 nm
     SAMPLE   Aspirin   0.21
                   Atropine   0.59
                   Bupivicaine   1.00 (7.5 min)
                   Caffeine   0.43
                   Codeine   0.16
                   Dropoperidol   2.15
                   Fentanyl   2.08
                   Lidocaine   0.43
                   Morphine   0.25
                   Scopolamine   0.24
                   Thiopental   0.79

     REFERENCE   A. N. Masoud, G. A. Scratchley, S. J. Stohs and
                   D. M. Wingard (1978) *J. Liq. Chromatog.* **1**, 607.

(e) COLUMN μBondapak C$_{18}$ 10 μm (30 cm × 4 mm i.d.)
     COLUMN TEMP
     MOBILE PHASE methanol–acetic acid–water (20 : 1 : 79)
     FLOW RATE   1 cm^3 min^{-1}
     PRESSURE   133 atm (2000 lbf in^{-2})
     DETECTOR   u.v. 280 nm
     SAMPLE   Theobromine   6.5 min
                   Caffeine   13.7 min

     REFERENCE   W. R. Kreiser and R. A. Martin (1978) *J. Assoc. Off.
                   Anal. Chem.* **61**, 1424.

8.2 Biochemicals

8.2.1 Amino Acids, Peptides and Proteins

(a) COLUMN   Pellosil HC (1 m × 4.0 mm i.d.)
COLUMN TEMP
MOBILE PHASE   dichloromethane + 1% methanol
FLOW RATE $2.5 \, cm^3 \, min^{-1}$
PRESSURE $350 \, lbf \, in^{-2}$
DETECTOR   u.v.
SAMPLE

N-Bz-Val Val-OMe	3.0 min
N-Bz-Val Gly-OMe	4.0 min
N-Bz-Gly Val-OMe	4.8 min

REFERENCE   R. S. Ward and A. Pelter (1974) *J. Chromatog. Sci.* **12**, 570.

(b) COLUMN μBondapak C_{18} (30 cm × 3.0 mm i.d.)
COLUMN TEMP Solvent A, pH 2.85 0.7% sodium decylsulphate + 0.5% acetic acid
MOBILE PHASE Solvent B 0.12% heptane sulphonate, 0.25% acetic acid, 20% acetonitrile, 10% methanol. Programme 2 min isocratic 4 min to 25% B 25% to 85% B over 20 min
FLOW RATE $1.5 \, cm^3 \, min^{-1}$
PRESSURE
DETECTOR   u.v. 340 nm
SAMPLE   as o-Phthalaldehyde-2-mercaptoethanol

Aspartate	4.0 min
Serine	4.5 min
Glutamate + Glycine	5.0 min
Threonine	5.5 min
Alanine	7.5 min
Cysteine	14.0 min
Valine	14.5 min
Methionine	15.5 min
Tyrosine	17.0 min
Isoleucine	19.0 min
Leucine	20.0 min
Phenylalanine	21.5 min
Histidine	23.5 min
Lysine	25.0 min
Arginine	26.5 min
H_2O	32.0 min

REFERENCE   M. K. Radjai and R. T. Hatch (1980) *J. Chromatog.* **196**, 319

(c) COLUMN μBondapak C_{18} (30 cm × 4 mm i.d.) 2 in tandem
COLUMN TEMP 25°C
MOBILE PHASE initial 30% methanol–70% $0.01 \, M$-K_2PO_4
                        final   70% methanol–30% $0.01 \, M$-K_2PO_4

FLOW RATE   1.2 cm^3 min^{-1}
PRESSURE
DETECTOR   365 nm fluorescence
SAMPLE   as Dansyl amino acids

Asparagine	5.6 min
Glutamate	6.0 min
Dansyl hydroxide	10.0 min
Serine	15.2 min
Threonine	16.4 min
Alanine	17.4 min
Proline	22.0 min
Valine	22.8 min
Methionine	23.2 min
Cystine	26.8 min
Isoleucine	26.8 min
Histidine	27.6 min
Arginine	27.6 min
Leucine	27.6 min
Lysine	28.8 min
Phenylalanine	28.8 min
Dansylamide	29.0 min

REFERENCE   R. Bongiovanni and W. Dutton (1978) *J. Liq. Chromatog.*
1, 617.

(d) COLUMN   μBondapak C$_{18}$ 10 μm (30 cm × 4 mm id.)
    COLUMN TEMP
    MOBILE PHASE water + 0.05% trifluoroacetic acid (pH 2.3)
    FLOW RATE   1.5 cm^3 min^{-1}
    PRESSURE
    DETECTOR   u.v.
    SAMPLE   L-Leu (Gly)$_3$ 5.4 min

    REFERENCE   C. A. Bishop, D. R. K. Harding, L. J. Meyer,
                     W. S. Hancock (1980) *J. Chromatog.* **192**, 222.

(e) COLUMN   Spherisorb ODS 5 μm (10 cm × 5 mm i.d.)
    COLUMN TEMP ambient
    MOBILE PHASE 50% methanol in 0.04 mole dm^{-3} disodium hydrogen
                     citrate buffer containing decyl sodium sulphate (pH 2.25)
    FLOW RATE   2.0 cm^3 min^{-1}
    PRESSURE
    DETECTOR   u.v. 280 nm
    SAMPLE

5-Hydroxy-3-indoleacetic acid	1.25 min
5-Hydroxytryptamine	2.0 min
5-Hydroxytryptophan	2.0 min
Kynurenine	2.25 min
Tryptophan	3.50 min

    REFERENCE   C. M. Riley, E. Tomlinson, T. M. Jeffries and
                     P. H. Redfern (1979) *J. Chromatog.* **162**, 153.

8.2.2   Carbohydrates

COLUMN	Pellosil HC (1 m x 4.0 mm i.d.)	
COLUMN TEMP		
MOBILE PHASE	dichloromethane	
FLOW RATE	2.5 cm^3 min^{-1}	
PRESSURE	200 lbf in^{-2}	
DETECTOR	u.v. 254 nm	
SAMPLE	β-Methyl D-allopyranoside 2,4,6-tribenzoate	4.75 min
	β-Methyl D-glucopyranoside 2,3,6-tribenzoate	7.25 min
	β-Methyl D-glucopyranoside 2,3,6-tribenzoate	8.75 min
	β-Methyl D-galactopyranoside 2,3,6-tribenzoate	10.0 min

REFERENCE   R. S. Ward and A. Pelter (1974) *J.Chromatog. Sci.* **12**. 570.

(b)

COLUMN	Porasil A (2 ft x 3/8 inch)	
COLUMN TEMP		
MOBILE PHASE	methyl ethyl ketone—water—acetone (85 : 10 : 5)	
FLOW RATE	2.1 cm^3 min^{-1}	
PRESSURE		
DETECTOR	Refractometer	
SAMPLE	β-D-galactose penta-acetate	10 min
	Phenyl β-D-glucopyranoside	22 min
	Methyl α-D-glucopyranoside	58 min

REFERENCE   G. P. Belue (1974) *J. Chromatog.* **100**, 233.

(c)

COLUMN	Bondapak AX/Corasil (1.24 m x 2 mm i.d.)	
COLUMN TEMP		
MOBILE PHASE	water—ethyl acetate—propan-2-ol	
FLOW RATE		
PRESSURE		
DETECTOR	refractometer	
SAMPLE	rhamnose	8.5 min
	xylose	10.0 min
	fructose	11.5 min
	dextrose	13.0 min
	sucrose	18.0 min
	lactose	27.0 min
	melibiose	27.0 min

REFERENCE   J. N. Little, D. F. Horgan, R. L. Cotter and R. V. Vivilecchia (1974) in *Bonded Stationary Phases in Chromatography* (ed. E. Grushka), Ann. Arbor. Science.

(d)   COLUMN   Bondapak-Carbohydrate (0.3 m x 4.0 mm i.d.)
      COLUMN TEMP
      MOBILE PHASE water—acetonitrile (15 : 85)
      FLOW RATE   2.0 cm^3 min^{-1}
      PRESSURE
      DETECTOR   refractometer

SAMPLE		
	Glycerol	3.4
	L-Rhamnose	3.6
	Ribose	4.0
	Xylose	4.3
	Fucose	4.4
	L-Arabinose	5.1
	Fructose	6.3
	Mannose	7.1
	Glucose	7.7
	Sucrose	15.0
	Maltose	19.4
	Melibiose	30.5

      REFERENCE   R. B. Meagher and A. Furst (1976) *J. Chromatog.* **117**, 211.

(e)   COLUMN   Aminex A 6 Li$^+$ (1 m x 4.0 mm i.d.)
      COLUMN TEMP 75°C
      MOBILE PHASE 85% ethanol—water
      FLOW RATE
      PRESSURE   1000 lbf in^{-2}
      DETECTOR   moving wire

SAMPLE		
	Rhamnose	20 min
	Xylose	33 min
	Arabinose	44 min
	Glucose	54 min
	Galactose	66 min
	Maltose	82 min

      REFERENCE   T. G. Lawrence (1975) *Chimia* **29**, 367:

(f)   COLUMN   Lichrosorb Si 60 5μm (25 cm × 2.1 mm i.d.)
      COLUMN TEMP 50°C
      MOBILE PHASE 0.1% water in acetonitrile
      FLOW RATE   0.2 cm^3 min^{-1}
      PRESSURE   100 bar
      DETECTOR   refractometer
      SAMPLE   20 μl aliquot from 6 g l^{-1} per sugar

	k'	
Fructose	4.3	
Glucose	5.1	
Mannose	5.7 (25 min)	

      REFERENCE   H. van Olst and G. E. H. Joosten (1979) *J. Liq. Chromatog.* **2**, 111

8.2.3   Lipids

(a) COLUMN   Vydac 35—44 μm (1.0 m x 2.0 mm i.d.)
    COLUMN TEMP 60°C
    MOBILE PHASE methanol—water (90 : 10)
    FLOW RATE   1.7 cm^3 min^{-1}
    PRESSURE   1500 lbf in^{-2}
    DETECTOR   refractometer
    SAMPLE   Methyl oleate   1.50 min
                  Methyl stearate   1.80 min
                  Methyl arachidate   2.30 min
                  Methyl behenate   3.20 min
                  Methyl lignocerate   5.30 min

    REFERENCE   P. T. S. Pei, R. S. Henly and S. Ramachandran (1975)
                  Lipids, **10**, 152.

(b) COLUMN   Bondapak C$_{18}$ Porasil (4 ft x 5/16 inch i.d.)
    COLUMN TEMP
    MOBILE PHASE 100% acetonitrile
    FLOW RATE   5.0 cm^3 min^{-1}
    PRESSURE
    DETECTOR   refractometer
    SAMPLE   Methyl linolenate   14.6 min
                  Methyl linoleate   18.4 min
                  Methyl oleate   24.8 min
                  Methyl palmitate   30.8 min
                  Methyl stearate   46.0 min

    REFERENCE   C. R. Schofield (1975) *Anal. Chem.* **47**, 1417.

(c) COLUMN   Corasil-C$_{18}$ (0.90 m x 0.17 mm i.d.)
    COLUMN TEMP
    MOBILE PHASE methanol—water (85 : 15)
    FLOW RATE   0.2 cm^3 min^{-1}
    PRESSURE   300 lbf in^{-2}
    DETECTOR   u.v. 254 nm
    SAMPLE   Naphthacyl linolenate   37 min
                  Naphthacyl linoleate   53 min
                  Naphthacyl oleate   85 min

    REFERENCE   M. W. Anders and M. J. Cooper (1974) *Anal. Chem.* **46**, 1849.

(d) COLUMN MicroPak Si-10 (0.50 m x 2.4 mm)
    COLUMN TEMP
    MOBILE PHASE hexane–dichloromethane–isopropanol(99.2 : 0.7 : 0.1)
    FLOW RATE 1.0 cm^3 min^{-1}
    PRESSURE
    DETECTOR u.v. 215 nm
    SAMPLE

	cis-methyl farnesoate	5.0 min
	trans-methyl farnesoate	7.6 min

    REFERENCE C. David Carr (1974) *Anal. Chem.* **46**, 743.

(e) COLUMN Adsorbosil-2 ADN (silver nitrate modified silica gel)
    (0.3 m x 9.3 mm i.d.)
    COLUMN TEMP
    MOBILE PHASE benzene
    FLOW RATE 5.5 cm^3 min^{-1}
    PRESSURE 2700 lbf in^{-2}
    DETECTOR refractometer
    SAMPLE

	dodec-9E-en-1-yl acetate	7.3 min
	dodec-9Z-en-1-yl acetate	15.5 min

    REFERENCE R. R. Heath, J. H. Tumlinson, R. E. Doolittle and A. T. Proveau (1975) *J. Chromatog. Sci.* **13**, 380.

(f) COLUMN ODS Sil-X-1 (0.5 m x 2.6 mm i.d.)
    COLUMN TEMP 50°C
    MOBILE PHASE acetonitrile–water (27 : 73) adjusted to pH 2.8 with phosphoric acid
    FLOW RATE 1.5 cm^3 min^{-1}
    PRESSURE
    DETECTOR u.v. 195 nm
    SAMPLE Prostaglandin $F_{2\alpha}$ 6 min

    REFERENCE Perkin-Elmer Instrument News, Vol. 24, Winter 1975.

(g) COLUMN Bondapak C_{18} (0.25 m x 4.0 mm i.d.)
    COLUMN TEMP
    MOBILE PHASE acetonitrile–water (50 : 50)
    FLOW RATE 1.2 cm^3 min^{-1}
    PRESSURE 1200 lbf in^{-2}
    DETECTOR u.v. 254 nm
    SAMPLE *p*-bromophenacyl esters of

	Prostaglandin F_2	11.5 min
	Prostaglandin E_2	14.0 min
	Prostaglandin D_2	16.0 min
	Prostaglandin A_2	32.0 min
	Prostaglandin B_2	34.0 min

    REFERENCE F. A. Fitzpatrick (1976) *Anal. Chem.* **48**, 499.

(h) COLUMN   Nucleosil 5 C_{18} Merck (25 cm × 4.6 mm i.d.)
COLUMN TEMP
MOBILE PHASE   water–acetonitrile–tetrahydrofuran (70 : 30 : 2)
FLOW RATE   1 cm^3 min^{-1}
PRESSURE   80 kg cm^{-2}
DETECTOR   u.v. 208 nm
SAMPLE   Prostaglandins (350 ng each)

PGF_1	2.0 min
PGF_2	3.5 min
PGE_2	4.0 min
PGE_1	4.6 min
PGA_2 } PGB_2 }	7.3 min
PGA_1 } PGB_1 }	9.5 min

REFERENCE   S. Inoyama, H. Hori, T. Shibata, Y. Ozawa, K. Yamagami, M. Imazu and H. Hayashida (1980) *J. Chromatog.* **194**, 85.

(i) COLUMN   Hitachi resin 2613
COLUMN TEMP   54°C
MOBILE PHASE   10% methylcellosolve–water
FLOW RATE   1 cm^3 min^{-1}
PRESSURE
DETECTOR   electrochemical (0.01M-p-benzoquinone-0.01M-hydroquinone-0.1M-KNO_3)
SAMPLE

Maleic acid	13 min
Fumaric acid	16 min
Formic acid	23 min
Acetic acid	27 min
Propionic acid	32 min
n-Butyric acid	36 min
Isovaleric acid	39 min
n-Valeric acid	45 min
Isocaproic acid	53 min
n-Caproic acid	60 min

REFERENCE   Y. Takata and G. Moto (1973) *Anal. Chem.* **45**, 1864.

(j) COLUMN   Lichrosorb C_{18} 10 μm (25 cm × 3 mm i.d.)
COLUMN TEMP   35°C
MOBILE PHASE   water–acetonitrile gradient of 38% acetonitrile to 75% acetonitrile
FLOW RATE   2 cm^3 min^{-1}
PRESSURE
DETECTOR   u.v. 254 nm
SAMPLE   as Naphthacyl ester

Lactic acid	2.62 min
Acetic acid	4.85 min
Propionic acid	7.16 min
Bromoacetonaphtol	7.80 min

Butyric acid	9.06 min
Valeric acid	10.18 min

REFERENCE   W. Distler (1980) *J. Chromatog.* **192**, 240

(k) COLUMN   Poragel 200 A (1 m × 4.0 mm i.d.)
COLUMN TEMP   45°C
MOBILE PHASE   5% aqueous acetone
FLOW RATE   0.66 cm^3 min^{-1}
PRESSURE   14 atm (210 lbf in^{-2})
DETECTOR   moving wire
SAMPLE

Monopalmitin	9 min
Palmitic acid	11 min
1,3-Dipalmitin	13 min
Cholesterol	18 min
Tripalmitin	24 min

REFERENCE   J. G. Lawrence (1973) *J. Chromatog.* **84**, 299.

(l) COLUMN   Spherisorb S 10W (0.30 m × 4.3 mm)
COLUMN TEMP
MOBILE PHASE   programmed: heptane—diethyl ether (99 : 1); chloroform—dioxan (92 : 8); heptane—diethyl ether—dioxan—propan-2-ol—water (20 : 20 : 30 : 30 : 1)
FLOW RATE   1.4 cm^3 min^{-1}
PRESSURE
DETECTOR   moving wire
SAMPLE

Squalene	2.52 min
Cetyl oleate	2.91 min
Methyl oleate	4.36 min
Triolein	5.50 min
Glycerol-l-oleate-2,3-diacetate	7.13 min
1,3-Diglyceride	8.00 min
1,2-Diglyceride + cholesterol	9.55 min
Mono-olein	13.42 min

REFERENCE   K. Aitzetmuller (1975) *J. Chromatog.* **13**, 454

(m) COLUMN   μBondapak C$_{18}$ (30 cm × 3.9 mm i.d.)
COLUMN TEMP
MOBILE PHASE   acetonitrile–acetone (42 : 58)
FLOW RATE   1.3 cm^3 min^{-1}
PRESSURE
DETECTOR   refractometer
SAMPLE   Triacylglycerols from cotton seed

REFERENCE   J. A. Bezard and M. A. Ouedraogo (1980) *J. Chromatog.* **196**, 279

(n) COLUMN   Partisil 5 5 μm (50 cm × 0.7 cm)
COLUMN TEMP
MOBILE PHASE   75% ethanol in hexane
FLOW RATE   4 cm^3 min^{-1}
PRESSURE
DETECTOR   205 nm
SAMPLE   Methyl 11-hydroperoxyoctadec-9c-enoate
                     Methyl 11-hydroperoxyoctadec-9t-enoate   18.0 min

REFERENCE   H. W. S-Chan and G. Levett (1977) *Chem. Ind.* 692.

(o) COLUMN   Lichrosorb RP-2 5 μm (5 cm × 4.6 mm i.d.)
COLUMN TEMP   40° C
MOBILE PHASE   Water (with 0.01 M-K_2HPO_4 + H_3PO_4 at pH 7.7)–
                     acetonitrile (90 : 10)
FLOW RATE   2 cm^3 min^{-1}
PRESSURE   26.6 atm (400 lbf in^{-2})
DETECTOR   u.v. at 403 nm and at 204 nm
SAMPLE   Neoxanthine   8.01 min
                     Violaxanthine   8.54 min
                     Luteine   9.34 min
                     Chlorophyll b   10.17 min
                     Chlorophyll a   10.46 min
                     Carotene   11.42 min
                     Phosphatidyl inositol   4.70 min
                     Phosphatidyl glycerol   4.87 min
                     Linolenic acid $\big\rbrace$
                     Linoleic acid   5.14 min
                     Oleic acid $\big\rbrace$
                     Stearic acid   5.53 min

REFERENCE   M. Riedmann and M. Tevini, Hewlett Packard Note
                     232–13

(p) COLUMN   Spherisorb ODS 5 μm (23 cm × 0.4 cm)
COLUMN TEMP
MOBILE PHASE   5.5% water in methanol
FLOW RATE   2 cm^3 min^{-1}
PRESSURE
DETECTOR   fluorescence and u.v. 254 nm
SAMPLE   Chlorophyll b   6.5 min
                     Chlorophyll a   9.5 min

REFERENCE   G. Liebezeit, (1980) *J. High Res. Chrom. Chrom. Comm.*
                     3, 531.

8.2.4 Nucleic Acid Constituents

(a) COLUMN   Perisorb KAT (30–40 μm) (0.50 m x 2.0 mm i.d.)
COLUMN TEMP
MOBILE PHASE 0.17M-formic acid + 0.35M-sodium chloride + 25%
              ammonia (pH 4.05)
FLOW RATE   0.8 cm^3 min^{-1}
PRESSURE   23 bar (335 lbf in^{-2})
DETECTOR   u.v. 254 nm
SAMPLE

Uracil	1.1 min
Hypoxanthine	1.9 min
Adenosine	3.5 min
Cytosine	8.8 min

REFERENCE   Merck, A. G., Darmstadt.

(b) COLUMN   Zipax SCX (0.5 m)
COLUMN TEMP
MOBILE PHASE 0.01M-HNO$_3$ + 0.05M-HN$_4$NO$_3$
FLOW RATE   1.5 cm^3 min^{-1}
PRESSURE   700 lbf in^{-2}
DETECTOR   u.v.
SAMPLE

Uracil	1.0 min
Cytidine	1.5 min
Guanine	2.2 min
Cytosine	4.2 min
Adenine	5.0 min

REFERENCE   Du Pont Co. Ltd., Instrument Division.

(c) COLUMN   10 μm Reverse Phase C$_{18}$
COLUMN TEMP
MOBILE PHASE sol a 0.02 mol l^{-1} KH$_2$PO$_4$ (pH 5.6)
FLOW RATE   sol b 60% methanol gradient slope 0.69% min (0–60%
              methanol in 87 min)
PRESSURE
DETECTOR   fluorescence
SAMPLE

Cytosine	2.28 min
Uridine diphosphoglucose	2.35 min
Uracil	4.13 min
Guanosine monophosphate	4.52 min
Thymine	9.30 min
Guanosine	14.40 min
+ 30 other nucleoside bases	

REFERENCE   R. A. Hartwick, S. P. Assenza and P. R. Brown (1979)
                  J. Chromatog. **186**, 647.

(g) COLUMN   Zipax SCX
     COLUMN TEMP   ambient
     MOBILE PHASE $0.05M$-NaH_2PO_4 + $0.05M$-KH_2PO_4 at pH 4.4
     FLOW RATE   1.2 cm^3 min^{-1}
     PRESSURE   1500 lbf in^{-2}
     DETECTOR   u.v.
     SAMPLE   Nicotinamide   1.75 min
                         Riboflavin   3.50 min
                         Pyridoxine   5.75 min

     REFERENCE   R. C. Williams, R. A. Henry and J. A. Schmit (1973)
                         J. Chromatog. Sci., 11, 358.

(h) COLUMN   Lichrosorb NH_2 10 μm (25 cm × 4.6 mm i.d.)
     COLUMN TEMP
     MOBILE PHASE   75% acetonitrile in 0.005 M-KH_2PO_4 (pH 4.4–4.7)
     FLOW RATE   3 cm^3 min^{-1}
     PRESSURE
     DETECTOR   u.v. 268 nm
     SAMPLE   Ascorbic acid   3.5 min
                         Isoascorbic acid 4.2 min

     REFERENCE   M. Huong Bui-Nguyen (1980) J. Chromatog. 196, 163.

8.2.6   Miscellaneous
(a) Carboxylic acids
     COLUMN   silica gel (6 μm) (300 m^2/g) (0.3 m x 3 mm i.d.)
     COLUMN TEMP
     MOBILE PHASE   0.1M-tetrabutylammonium hydrogen sulphate–
                         0.11M-Na_2HPO_4–0.065M-Na_3PO_4
     FLOW RATE   54 cm^3 min^{-1}
     PRESSURE   2200 lbf in^{-2}
     DETECTOR   u.v.
     SAMPLE   Toluene   0.59 min
                         Indoleacetic acid   0.90 min
                         Homovanillic acid   1.94 min
                         Vanilmandelic acid   2.43 min
                         5-Hydroxyindole-3-acetic acid   3.10 min
                         as their ion-pairs with tetrabutylammonium ions

     REFERENCE   B. A. Persson and B. L. Karger (1974) J. Chromatog. Sci.
                         12, 521.

(b) Urine extracts

COLUMN   LiChrosorb SI 100 (10 M) + 0.1 M-tetrabutylammonium
acetate in Tris buffer (pH 8.3)

COLUMN TEMP

MOBILE PHASE butan-1-ol—dichloromethane—n-hexane (3 : 8 : 9)

FLOW RATE

PRESSURE

DETECTOR   u.v.

SAMPLE   5-Hydroxyindole-3-acetic acid 5.0 min

REFERENCE   B. A. Persson and P. O. Lagerstrom (1976) *J. Chromatog. Sci.* **122**, 305.

(c) Tricarboxylic acid

COLUMN   Corasil II (1.0 m x 2. 1 mm i.d.)

COLUMN TEMP

MOBILE PHASE isopropanol—n-hexane (1 : 50)

FLOW RATE   1.12 cm^3 min^{-1}

PRESSURE

DETECTOR   u.v. 210 nm

SAMPLE

Oxalic acid	3 min
Lactic acid	6 min
α-Ketoglutaric acid	12 min
trans-Aconitic acid	16 min

REFERENCE   W. Funasaka, T. Hanai, and K. Fujimura (1975)
J. Chromatog. Sci **12**, 517.

(d) Tricarboxylic acids

COLUMN   Silica gel 25 μm (0.3 m × 2.3 mm i.d.)

COLUMN TEMP 19°C

MOBILE PHASE 6% t-amyl alcohol + 50 cm^3 0.1M-$(NH_4)_2SO_4$ + 50 cm^3
chloroform

PRESSURE

DETECTOR   u.v. 432 nm

SAMPLE

Acetic acid	11.6
Fumaric acid	12.8
Pyruvic acid	18.1
Glutaric acid	20.6
Lactic acid	45.0
Succinic acid	45.0
α-Ketoglutaric acid	90.0

REFERENCE   K. W. Stahl, G. Schafer and W. Lamprecht (1972)
J. Chromatog. Sci. **10**, 95.

(e) Amines

COLUMN   Porasil E (300 m^2/g) + 0.1M-HClO$_4$ —0.9 M-NaClO$_4$
(0.3 m × 3.0 mm i.d.)

COLUMN TEMP

MOBILE PHASE ethyl acetate–tributyl phosphate–hexane (72.5 : 10 : 17.5)
FLOW RATE 48 cm^3 min^{-1}
PRESSURE
DETECTOR u.v.
SAMPLE

Phenethylamine	0.6	
Tyramine	1.0 (0.9 min)	
3-Methoxytyramine	1.6	
Dopamine	2.0	
Normetanephrine	3.3	
Metanephrine	4.0	
Noradrenaline	4.8	
Adrenaline	5.3	

REFERENCE B. A. Persson and B. L. Karger (1974) *J. Chromatog. Sci.* **12**, 521.

(f) Plant growth factors
COLUMN μBondapak C$_{18}$ (25 cm × 6.4 mm i.d.)
COLUMN TEMP
MOBILE PHASE 12.5% methanol (pH 2.8) for 20 min then linear to 50% methanol (pH 2.8) over 30 min
FLOW RATE 3 cm^3 min^{-1}
PRESSURE
DETECTOR u.v. 280 nm
SAMPLE

trans-Zeatin	14 min
Zeatin riboside	32 min
Indoleacetic acid	40 min
Abscisic acid	51 min

REFERENCE R. N. Arteca, B. W. Pooviah and O. E. Smith (1980) *Pl. Physiol.* **65**, 1216

(g) COLUMN Bondapak C$_{18}$ Porasil B (60 cm × 6.5 mm. i.d.)
COLUMN TEMP
MOBILE PHASE solvent A 1% aqueous acetic acid
solvent B 95% ethanol
gradient of B 30 to 100% in 25 min
FLOW RATE 9.9 cm^3 min^{-1}
PRESSURE
DETECTOR
SAMPLE

Fractions collected every minute
Gibberellins

Gibberellin A$_8$	10–12 min
Gibberellin A$_{29}$	11–12 min
Gibberellin A$_{23}$	11–13 min
Gibberellin A$_1$	16–18 min

other Gibberellins analysed were A$_5$, A$_{44}$, A$_{19}$, A$_{13}$, A$_{17}$, A$_{36}$, A$_{37}$, A$_4$, A$_7$, A$_{14}$, A$_9$, A$_{25}$

REFERENCE M. G. Jones, J. D. Metzgar and J. A. D. Zeevaart (1980) *Pl. Physiol.* **65**, 218.

8.3 Food chemicals

(a) COLUMN   Aminex A 25 (0.9 m x 1/4 inch o.d.)
COLUMN TEMP $70°$
MOBILE PHASE 1.0M-sodium formate
FLOW RATE cm^3/min
PRESSURE
DETECTOR   refractometer
SAMPLE   Grape juice:
                                  Citric acid   27.2 min
                                  Malic acid   30.0 min
                                  Tartaric acid   33.3 min

REFERENCE   J. K. Palmer and D. M. List (1973) *J. Agric. Food Chem.* 21, 903.

(b) Food contaminants
COLUMN   Zorbax SIL (0.25 m x 2.1 mm i.d.)
COLUMN TEMP ambient
MOBILE PHASE 60% CH_2Cl_2 (50% H_2O sat.)—40% $CHCl_3$
                       (50% H_2O sat.)—0.1% methanol
FLOW RATE   0.7 cm^3 min^{-1}
PRESSURE   1500 lbf in^{-2}
DETECTOR   u.v. 254 nm
                           365 nm
SAMPLE   Aflatoxins
                            B_1 8.0 min
                            G_1 9.5 min
                            B_2 11.0 min
                            G_2 13.5 min

REFERENCE   Du Pont Co. Ltd., Instrument Products Division.

(c) Food contaminants
COLUMN μBondapak C_{18} 10 μm (30 cm x 4.0 mm i.d.)
COLUMN TEMP 35°C
MOBILE PHASE methanol–water (40 : 60)
FLOW RATE   2 cm^3 min^{-1}
PRESSURE   150–330 bar
DETECTOR   fluorescence
SAMPLE   Aflatoxin G_2 4.54 min
                          Aflatoxin G_1 5.58 min
                          Aflatoxin B_2 6.82 min
                          Aflatoxin B_1 8.56 min

REFERENCE   R. Knutti, Ch. Balsiger and K. Sutter (1979) *Chromatographia* 12, 349.

(d) Frying fat components

COLUMN   Merckogel SI 50 (36−75 μm) (0.2 m x 4 mm i.d.)

COLUMN TEMP

MOBILE PHASE heptane; heptane−isopropyl ether (80 : 20);
heptane−isopropyl ether−ethanol−water (20 : 30 : 50 : 1)

FLOW RATE

PRESSURE

DETECTOR   moving wire

SAMPLE

Sterol esters	2.1 min
Triglycerides	3.2 min
Dimer triglycerides	6.6 min

REFERENCE   K. Aitzetmuller and G. Guhr (1976) *Fette, Seifen, Anstrichm.* **78**, 83.

(e) Food additives

COLUMN   Zipax SAX (1.0 m × 4.0 mm i.d.)

COLUMN TEMP

MOBILE PHASE 0.01M-sodium tetraborate + 0.02M-sodium nitrate

FLOW RATE   1.8 cm^3 min^{-1}

PRESSURE   350 lbf in^{-2}

DETECTOR   u.v.

SAMPLE

Benzoate	4.1 min
Saccharin	13.5 min

REFERENCE (Jobling) J. J. Nelson (1973) *J. Chromatog. Sci.* **11**, 28.

(f) Food additives

COLUMN   Micropak NH$_2$ (0.25 m x 2 mm i.d.)

COLUMN TEMP

MOBILE PHASE isopropanol−hexane (30 : 70)

FLOW RATE   1 cm^3 min^{-1}

PRESSURE

DETECTOR   u.v. 210 nm

SAMPLE   Lecithin   1.6 min

REFERENCE   Varian Associates.

(g) Food additives

COLUMN   Whatman PXS 10/ 25 SCX (25 cm × 4.6 mm i.d.)

COLUMN TEMP

MOBILE PHASE acetonitrile−methanol−water (400 : 100 : 4)

FLOW RATE   2.5 cm^3 min^{-1}

PRESSURE

DETECTOR   u.v. 203 nm

SAMPLE

Phosphatidylethanolamine	4.5 min
Lysophosphatidylethanolamine	7.0 min
Phosphatidylcholine	10.0 min

REFERENCE   R. W. Gross and B. E. Sobel (1980) *J. Chromatog.* **197**, 79.

(h) Food additives

COLUMN	μPorasil (30 cm $\times$ 4.8 mm i.d.)
COLUMN TEMP	
MOBILE PHASE	acetonitrile–methanol–water (65 : 21 : 14)
FLOW RATE	2.0 cm^3 min^{-1}
PRESSURE	
DETECTOR	refractometer and u.v. 210 nm
SAMPLE	Phosphatidylcholine 7.0 min

REFERENCE   W. J. Hurst and R. A. Martin (1980) *J. Am. Oil Chem. Soc.* 57, 307.

(i) Antioxidants

COLUMN	Micropak Al-5 (0.15 m x 2 mm i.d.)
COLUMN TEMP	
MOBILE PHASE	programme: hexane + dichloromethane
FLOW RATE	1 cm^3 min^{-1}
PRESSURE	
DETECTOR	u.v.
SAMPLE	Butylated hydroxytoluene 0.45 min
	Triphenyl phosphate 1.23 min
	Butylated hydroxyanisole 3.00 min

REFERENCE   Varian Associates.

8.4   Heavy Industrial Chemicals

8.4.1   Pesticides

(a)

COLUMN	Perisorb (30–40 μm) (0.5 m x 2.0 mm i.d.)
COLUMN TEMP	
MOBILE PHASE	n-heptane–ethyl acetate (95 : 5)
FLOW RATE	1 cm^3 min^{-1}
PRESSURE	10 bar (145 lbf in^{-2})
DETECTOR	u.v. 280 nm
SAMPLE	Methyl monochlorophenoxyacetate 1.3 min
	Methyl 2,4-dichlorophenoxyacetate 1.8 min

REFERENCE   Merck, A. G., Darmstadt.

(b)

COLUMN	μBondapak C$_{18}$ (30 cm $\times$ 0.4 cm i.d.)
COLUMN TEMP	ambient
MOBILE PHASE	100% 0.01 M-NaH$_2$PO$_4$–Na$_2$HPO$_4$ (pH 7.0) to 100% methanol in 30 min
FLOW RATE	2.0 cm^3 min^{-1}
PRESSURE	
DETECTOR	280 nm

SAMPLE	3-Hydroxy-2,4-dichlorophenoxyacetic acid	7.0 min
	2-Chloro-4-hydroxyphenoxyacetic acid	8.0 min
	5-Hydroxy-2,4-dichlorophenoxyacetic acid	13.0 min
	2,4-Dichlorophenoxyacetic acid	16.0 min

REFERENCE   A. D. Drinkwine, D. W. Bristol, J. R. Fleeker (1979)
J. Chromatog. **174**, 264.

(c) COLUMN   Micropak NH_2 (0.25 m x 2.2 mm i.d.)
   COLUMN TEMP
   MOBILE PHASE   solvent programme A: iso-octane
                  B: 33% isopropanol in CH_2Cl_2
   FLOW RATE   2 cm^3 min^{-1}
   PRESSURE
   DETECTOR   u.v. 254 nm
   SAMPLE   Diuron   1.25
                  Secondary amine degradation product   2.50
                  Primary amine degradation product   4.20

   REFERENCE   Varian Associates.

(d) COLUMN   Micropak SI 10 (0.5 m x 2 mm i.d.)
   COLUMN TEMP
   MOBILE PHASE   solvent programme A: 99.9% hexane + 0.1% isopropanol
                  B: 90% methylene chloride + 10%
                     isopropanol
   FLOW RATE   2 cm^3 min^{-1}
   PRESSURE
   DETECTOR
   SAMPLE   Propazine   6.1 min
                  Atrazine   7.0 min
                  Simazine   8.1 min
                  Cotoran   13.0 min

$R^1 = R^2 = H$ = simazine
$R^1 = CH_3$, $R^2 = H$ = atrazine
$R^1 = CH_3$, $R^2 = CH_3$ = propazine

cotoran

REFERENCE   Varian Associates.

(e) COLUMN   Corasil (1.0 m × 2.4 mm i.d.)
COLUMN TEMP
MOBILE PHASE 2% acetone in hexane
FLOW RATE
PRESSURE
DETECTOR
SAMPLE   dansyl derivatives of:

Butacarb	1.0 min
Mesurol	1.0 min
Carbofuran	2.0 min
Carzol	3.0 min
Mobam	4.0 min
Dioxacarb	13.0 min

REFERENCE   R. W. Frei, J. F. Lawrence, J. Hope and R. M. Cassidy (1974) *J. Chromatog. Sci.* 12, 40.

(f) COLUMN   Zipax TM G (1 m x 2.1 mm i.d.)
COLUMN TEMP ambient
MOBILE PHASE n-hexane saturated with trimethylene glycol
FLOW RATE   2 cm^3 min^{-1}
PRESSURE   1200 lbf in^{-2}
DETECTOR   u.v.
SAMPLE

Carbaryl	3.20 min
1-naphthol	8.20 min

REFERENCE   Du Pont Co. Ltd., Instrument Products Division.

(g) COLUMN   Corasil/C$_{18}$ 37–50 μm (4 ft x 2.0 mm i.d.)
COLUMN TEMP
MOBILE PHASE 0.05M-KCl in 30% methanol–70% water
FLOW RATE   0.5 cm^3 min^{-1}
PRESSURE   500 lbf in^{-2}
DETECTOR   polarographic
SAMPLE

p-Nitrophenol	2.5 min
Methyl parathion	5.0 min
Parathion	12.0 min

REFERENCE   R. Stillman and T. S. Ma (1974) *Mikrochimica Acta*, 641.

(h) COLUMN   Silanized kieselguhr (28–40 μm) + 10% 2,2,4-trimethyl-pentane (0.20 m x 2.75 mm i.d.)
COLUMN TEMP
MOBILE PHASE water–ethanol–acetic acid–sodium hydroxide–potassium chloride (60.1 : 38.8 : 0.80 : 0.21 : 0.09 w/w)
FLOW RATE   1 cm^3 min^{-1}
PRESSURE
DETECTOR   polarographic

SAMPLE	p-Nitrophenol	0.84 min
	Methyl parathion	1.45 min
	Oxygen	2.46 min
	Parathion	4.15 min

REFERENCE   J. G. Koen, J. F. K. Huber, H. Poppe and G. den Boef
(1970) *J. Chromatog. Sci.* **8**, 192.

(i) COLUMN   Permaphase ETH
    COLUMN TEMP
    MOBILE PHASE n-hexane
    FLOW RATE $1 \text{ cm}^3 \text{ min}^{-1}$
    PRESSURE 300 lbf in^{-2}
    DETECTOR   u.v.
    SAMPLE   Abate lavicide   4.0 min

REFERENCE   (Du Pont) R. A. Henry, J. A. Schmit and J. F. Dieckman
(1971) *Anal Chem.* **43**, 7.

(j) COLUMN   Spherosil XOA + oxydipropionitrile 10−20
                    (0.25 m x 4 mm i.d.)
    COLUMN TEMP
    MOBILE PHASE hexane
    FLOW RATE
    PRESSURE   60 bar (870 lbf in^{-2})
    DETECTOR   u.v.
    SAMPLE   Aldrin   0.9 min
                 Heptachlor   1.0 min
                 D.D.T.   1.2 min
                 Dieldrin   3.3 min

REFERENCE   J. Vermont, M. Deleuil, A. J. De Vries, and C. L. Guillemin
(1975) *Anal. Chem.* **47**, 1329.

(k) COLUMN   MicroPak + 33% oxydipropionitrile (0.5 m x 2 mm i.d.)
    COLUMN TEMP
    MOBILE PHASE iso-octane
    FLOW RATE $0.5 \text{ cm}^3 \text{ min}^{-1}$
    PRESSURE 375 lbf in^{-2}
    DETECTOR   u.v. 254 min
    SAMPLE   Mirex   1.9 min
                 Aldrin   2.4 min
                 DDT   4.1 min
                 Dieldrin   4.8 min
                 TDE   8.1 min
                 Methoxychlor 13.7 min

REFERENCE   Varian Associates.

(l) COLUMN   Lichrosorb Si 60 5 μm (15 cm × 4.7 mm i.d.)
    COLUMN TEMP
    MOBILE PHASE *n*-hexane–di-isopropyl ether (93 : 7)
    FLOW RATE   80 cm^3 h^{-1}
    PRESSURE   33.3 atm (500 lbf in^{-2})
    DETECTOR   u.v. 230 nm
    SAMPLE   Diphenylamine   4.7 min
                      Decamethrin (pyrethroid)   7.6 min
                      Cyano-3-phenoxybenzyl-*cis*-1R, 3R, 2,2 dimethyl-3-
                          (2,2-dibromovinyl) cyclopropane carboxylate

    REFERENCE   D. Mourot, B. Delepine, J. Boisseau and G. Gayot (1979)
                  J. Chromatog. **173**, 412

(m) COLUMN   Lichrosorb RP- 8 10 μm (25 cm × 3.2 mm i.d.)
    COLUMN TEMP
    MOBILE PHASE methanol–water (60 : 40)
    FLOW RATE   1 cm^3 min^{-1}
    PRESSURE
    DETECTOR   u.v. 238 nm
    SAMPLE   Barban herbicide   12.0 min
                      (4-chlorobut-2-ynyl (*N*-3-chlorophenyl) carbamate)

    REFERENCE   J. F. Lawrence, L. G. Panopio and H. A. McLeod (1980)
                  J. Chromatog. **195**, 113.

8.4.2   *Petroleum Products*

(a) COLUMN   Perisorb A (30–40 μm) + 1.3% 3,3-oxydipropionitrile
                  (0.5 m x 2.0 mm i.d.)
    COLUMN TEMP
    MOBILE PHASE heptane saturated with 3,3-oxydipropionitrile
    FLOW RATE   50 cm^3 min^{-1}
    PRESSURE   74 bar (1070 lbf in^{-2})
    DETECTOR   u.v. 254 nm
    SAMPLE   Dibutyl phthalate   6 s
                      Diethyl phthalate   8 s
                      Dimethyl phthalate   17 s

    REFERENCE   Merck

(b) COLUMN   Merck SI 60 (40 μm) + added water (0.45 m x 4.0 mm i.d.)
    COLUMN TEMP
    MOBILE PHASE n-hexane
    FLOW RATE   2.4 cm^3 min^{-1}
    PRESSURE   400 lbf in^{-2}
    DETECTOR   u.v.
    SAMPLE   Benzene   4.0 min
                      Naphthalene   5.6 min
                      Phenanthrene   9.6 min

    REFERENCE   Jobling (Corning) Laboratory Division.

(c) COLUMN   Durapak OPN (2 ft × 4 mm i.d.)
    COLUMN TEMP
    MOBILE PHASE iso-octane
    FLOW RATE $1.0\ cm^3\ min^{-1}$
    PRESSURE $20\ lbf\ in^{-2}$
    DETECTOR   u.v.
    SAMPLE

Anthracene	0.58 min
Pyrene	1.00 min
Chrysene	1.73 min
Perylene	3.67 min
Anthanthrene	5.00 min

REFERENCE   N. F. Ives and L. Giuffrida (1972) *J. Assoc. Off. Anal. Chem.* **55**, 757.

(d) COLUMN   Cellulose acetate (2 ft × 4.0 mm i.d.)
    COLUMN TEMP
    MOBILE PHASE methanol
    FLOW RATE $0.015\ cm^3\ min^{-1}$
    PRESSURE $40\ lbf\ in^{-2}$
    DETECTOR   u.v. 335 nm and 363 nm
    SAMPLE

Anthracene	9 min
Benzo[e]pyrene	11 min
Pyrene	20 min
Benzo[a]pyrene	38 min

REFERENCE   N. F. Ives and L. Giuffrida (1972) *J. Assoc. Off. Anal. Chem.* **55**, 757.

(e) COLUMN   Merckosorb SI 60 (30 μm) + 50% Fractonitril III
                  (2.5 m × 2.0 mm i.d.)
    COLUMN TEMP
    MOBILE PHASE n-heptane saturated with Fractonitril III
    FLOW RATE $15\ cm^3\ h^{-1}$
    PRESSURE
    DETECTOR   u.v. 254 nm
    SAMPLE

Benzene	21 min
Naphthalene	28 min
Anthracene	35 min
Pyrene	42 min
Fluoranthene	47 min
Tetracene	51 min
Chrysene	58 min
Benzpyrene	70 min
Coronene	88 min

REFERENCE   Merck A. G., Darmstadt.

(f) COLUMN   Perisorb A (30—40 μm) (0.5 m × 2.0 mm i.d.)
     COLUMN TEMP
     MOBILE PHASE n-heptane
     FLOW RATE   5 cm^3 min^{-1}
     PRESSURE
     DETECTOR   u.v. 254 nm
     SAMPLE

Benzene	9 s
Biphenyl	11 s
m-Terphenyl	18 s
m-Quaterphenyl	34 s
m-Quinquephenyl	74 s

     REFERENCE   Merck, A. G. Darmstadt.

(g) COLUMN   Lichrosorb RP-8 5 μm (20 cm × 4.6 mm i.d.)
     COLUMN TEMP
     MOBILE PHASE water—methanol (20 : 80)
     FLOW RATE   1 cm^3 min^{-1}
     PRESSURE
     DETECTOR   spectrofluorimeter
     SAMPLE

Phenanthrene	11.0 min
Pyrene	15.0 min
Chrysene	17.0 min
Benzo[α] pyrene	19.0 min

     REFERENCE   J. M. Colin and G. Vion (1980) *Analusis* 8, 224.

(h) COLUMN   Jasco SS-05-254 (25 cm × 4.6 mm i.d.) + Shodex Silica
                       Pack — 254 (25 cm × 4.6 mm i.d.)
     COLUMN TEMP 273 K
     MOBILE PHASE n-hexane
     FLOW RATE   0.4 cm^3 min^{-1}
     PRESSURE
     DETECTOR   u.v. 254 nm + refractometer
     SAMPLE

Methylcyclohexane	4.7 min
Oct-l-ene	5.0 min
Octa-1,7-diene	6.0 min
Penta-1,3-diene	7.5 min
Benzene	8.6 min
m-Xylene	10.5 min
Anthracene	19.0 min

     REFERENCE   K. Jinno, H. Nomura and Y. Hirata (1980) *J. High Res. Chrom., Chrom. Comm.* 3, 503.

8.4.3 Pollutants

     COLUMN   Zipax SCX (1.0 m × 2.1 mm i.d.)
     COLUMN TEMP 50°C
     MOBILE PHASE 0.2M-KH$_2$PO$_4$
     FLOW RATE   1.3 cm^3 min^{-1}
     PRESSURE   1200 lbf in^{-2}

DETECTOR u.v.
SAMPLE Aniline in waste stream 3.5 min

REFERENCE J. J. Kirkland (1974) *Analyst* **99**, 859.

8.4.4 Polymers

(a) COLUMN Three in series: CPG 75, CPG 240, CPG 700
(4 ft x 2 mm i.d.)
COLUMN TEMP
MOBILE PHASE dichloromethane
FLOW RATE 5.0 cm^3 min^{-1}
PRESSURE 8500 lbf in^{-2}
DETECTOR u.v. 254 nm
SAMPLE polystyrene: mol. wt. 1 800 000 0.75 min
mol. wt. 92 000 0.90 min
mol. wt. 4 000 1.25 min

REFERENCE Varian Associates.

(b) COLUMN porous silica microspheres (0.25 m x 2.1 mm i.d.)
COLUMN TEMP 60°C
MOBILE PHASE tetrahydrofuran
FLOW RATE 1.0 cm^3 min^{-1}
PRESSURE 1625 lbf in^{-2}
DETECTOR u.v.
SAMPLE polystyrene fractions;
mol. wt. 2 030 33.5 s
mol. wt. 51 000 26.0 s
mol. wt. 411 000 20.0 s

REFERENCE J. J. Kirkland (1972) *J. Chromatog. Sci.* **10**, 593.

8.4.5 Surfactants

(a) COLUMN Micropak NH$_2$ (0.25 m x 2 mm i.d.)
COLUMN TEMP
MOBILE PHASE non-linear gradient 0 to 100% isopropanol in hexane
FLOW RATE 1 cm^3 min^{-1}
PRESSURE
DETECTOR u.v. 220 nm
SAMPLE non-ionics surfactant Igepal 430

$n = 2$ 6.0 min
$n = 3$ 9.0 min
$n = 4$ 12.0 min

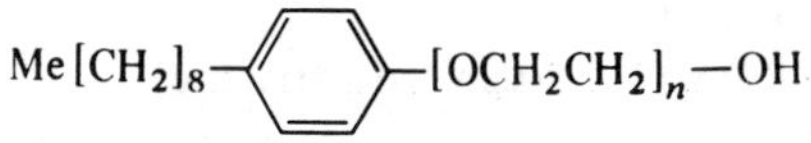

$$Me[CH_2]_8 - \langle \rangle - [OCH_2CH_2]_n - OH$$

REFERENCE Varian Associates.

(b) COLUMN   Lichrosorb RP-8 5 μm (15 cm × 6.35 mm i.d.)
    COLUMN TEMP
    MOBILE PHASE water–methanol (85 : 15) + cetrimide 0.5 g/100 cm^3
    FLOW RATE
    PRESSURE
    DETECTOR   u.v. 254 nm + refractometer
    SAMPLE   Alkylbenzene sulphonates
                   C alkyl   4.4 min
                   C alkyl   5.3 min
                   C alkyl   6.6 min
                   C alkyl   8.4 min
                   C alkyl   10.6 min
                   C alkyl   14.0 min

    REFERENCE   D. Thomas and J. L. Rocca (1979) *Analusis* **7**, 386.

(c) COLUMN   Lichrosorb RP-8 5 μm (15 cm × 6.35 mm)
    COLUMN TEMP
    MOBILE PHASE methanol–water (88 : 12, v/v) (pH 9.5)
    FLOW RATE
    PRESSURE
    DETECTOR   refractometer
    SAMPLE   Non-ionic detergent – average ethylene oxide content
                   Fatty acid chain length C_{10} 4.25 min
                                           C_{12} 5.5 min
                                           C_{14} 6.5 min
                                           C_{16} 8.5 min
                                           C_{18} 13.5 min

    REFERENCE   D. Thomas and J. L. Rocca (1979) *Analusis* **7**, 386.

8.4.6   *Explosives and Propellants*
(a) COLUMN   Corasil II (1.0 m x 3.0 mm i.d.)
    COLUMN TEMP
    MOBILE PHASE 30% dioxan–70% cyclohexane
    FLOW RATE   0.6 cm^3 min^{-1}
    PRESSURE   190 lbf in^{-2}
    DETECTOR   u.v.
    SAMPLE   TNT (2,4,6-trinitrotoluene)   6 min
                   Tetryl   7 min
                   RDX (1,3,5-trinitro-*s*-triazine)   21 min

    REFERENCE   J. O. Doali and A. A. Juhasz (1974) *J. Chromatog. Sci.*
                   12, 51.

(b) COLUMN   Vydac (30–44 μm) (1.0 m x 2.1 mm i.d.)
     COLUMN TEMP
     MOBILE PHASE   dichloromethane–hexane (60 : 40)
     FLOW RATE   0.75 cm^3 min^{-1}
     PRESSURE   450 lbf in^{-2}
     DETECTOR
     SAMPLE   Trinitroglycerine   2.5 min
                     1,3-Dinitroglycerine   6.5 min
                     1,2-Dinitroglycerine   8.0 min

     REFERENCE   C. D. Chandler, G. R. Gibson and W. T. Bolleter (1974)
                     J. Chromatog. **100**, 185.

(c) COLUMN   Vydac 30–44 μm (1.0 m x 2.1 mm i.d.)
     COLUMN TEMP
     MOBILE PHASE   1,1-dichloroethane
     FLOW RATE   0.8 cm^3 min^{-1}
     PRESSURE   450 lbf in^{-2}
     DETECTOR   u.v.
     SAMPLE   Nitroglycerine   2.4 min
                     Diethyl phthalate   4.6 min
                     Ethyl centralite   13.6 min
                     Acetanilide   20.6 min

     REFERENCE   R. W. Dalton, C. D. Chandler and W. T. Bolleter (1975)
                     J. Chromatog. Sci. **13**, 40

(d) COLUMN   Lichrosorb RP-8 10 μm (25 cm x 4.6 mm i.d.)
     COLUMN TEMP   40°C
     MOBILE PHASE   methanol–water (50 : 50, v/v) containing 5×10^3 M-
                     tetrabutylammonium phosphate (pH 7.5)
     FLOW RATE   2.0 cm^3 min^{-1}
     PRESSURE
     DETECTOR   u.v. 254 nm
     SAMPLE   Water   1.5 min
                     2-Amino-4,6-dinitrophenol   3.06 min
                     2,4-Dinitrophenol   4.26 min
                     2-Methyl-4,6-dinitrophenol   6.34 min
                     2,4,6-Trinitrophenol   7.67 min
                     3-Methyl-2,4,6-trinitrophenol 11.50 min

     REFERENCE   C. Hoffsommer, D. J. Glover and C. Y. Hazzard (1980)
                     J. Chromatog. **195**, 435.

(e) COLUMN μBondapak C_{18}
     COLUMN TEMP
     MOBILE PHASE 60% methanol in water
     FLOW RATE   2 cm^3 min^{-1}
     PRESSURE
     DETECTOR   u.v. 200 nm
     SAMPLE   Nitroglycerine   4 min

     REFERENCE   W. C. Crouthamel and B. Dorsch (1979) *J. Pharm. Sci.* **68**, 237

(f) COLUMN   Bondapak C_{18}
     COLUMN TEMP
     MOBILE PHASE 40% methanol in water
     FLOW RATE   100 cm^3 h^{-1}
     PRESSURE
     DETECTOR   230 nm
     SAMPLE   HMX (1,3,5,7-tetranitro-1,3,5,7-tetraazacyclo-octane)

Sample	Time
HMX (1,3,5,7-tetranitro-1,3,5,7-tetraazacyclo-octane)	3.5 min
RDX (1,3,5-trinitro-1,3,5-triazacyclohexane)	5.0 min
m-Nitrophenol	7.0 min
TNT (2,4,6-trinitrotoluene)	10.2 min

     REFERENCE   E. P. Meier, L. G. Taft, A. P. Graffeo and T. B. Stanford (1977) *Proc. 4th Joint Conf. Sensing of Environmental Pollutants*, New Orleans, Nov., 1977, p. 487.

8.4.7 Dyestuffs

(a) COLUMN   Zipax/1% oxydipropionitrile (1.0 m × 2.1 mm i.d.)
     COLUMN TEMP
     MOBILE PHASE hexane
     FLOW RATE   1.50 cm^3 min^{-1}
     PRESSURE   500–1000 lbf in^{-2}
     DETECTOR   u.v. 254 min
     SAMPLE   Azo dyes

Sample	Time
Du Pont Oil Red	1 min
CI Disperse Red 65	7 min
CI Disperse Orange 3	12 min
CI Disperse Yellow 3	18 min

     REFERENCE   R. J. Passarelli and E. S. Jacobs (1975) *J. Chromatog. Sci.* **13**, 153.

(b) COLUMN   Zipax/HCP (1 m × 2.1 mm i.d.)
     COLUMN TEMP 50°C
     MOBILE PHASE ethanol–water (50 : 50)
     FLOW RATE   0.46 cm^3 min^{-1}
     PRESSURE   500–1000 lbf in^{-2}

DETECTOR   u.v. 254 nm
SAMPLE

C.I. Disperse Red II	3.5 min
C.I. Solvent Blue II	7.0 min
1-Hydroxy-4-anilinoanthraquinone	9.0 min
C.I. Solvent Violet 13	13.5 min

REFERENCE   R. J. Passarelli and E. S. Jacobs (1975) *J. Chromatog. Sci.* **13**, 153.

(c) COLUMN   Micropak SI-10 (0.15 m x 2.1 mm)
COLUMN TEMP
MOBILE PHASE 10% dichloromethane—hexane
FLOW RATE $132\ cm^3\ h^{-1}$
PRESSURE $350\ lbf\ in^{-2}$
DETECTOR   u.v.
SAMPLE   Compounds shown in Fig. 1.1

1. 0.15 min
2. 0.30 min
3. 0.50 min
4. 1.05 min
5. 1.30 min
6. 4.60 min

REFERENCE   R. E. Majors (1973) *Anal. Chem.* **45**, 755.

(d) COLUMN   MicroPak Al-5 (0.15 m x 2.0 mm i.d.)
COLUMN TEMP
MOBILE PHASE 1% dichloromethane—hexane
FLOW RATE $2\ cm^3\ min^{-1}$
PRESSURE $1350\ lbf\ in^{-2}$
DETECTOR   u.v.
SAMPLE

Azobenzene	0.38 min
N,N-Diethyl-*p*-aminoazobenzene	0.44 min
N-Ethyl-*p*-aminoazobenzene	0.69 min
p-Aminoazobenzene	2.38 min

REFERENCE   Varian Associates.

8.5 Inorganic

(a) COLUMN   Resin AG WX 12 25 μm (100 cm × 7 mm i.d.)
COLUMN TEMP 80°C
MOBILE PHASE 0.2M-ammonium hydroxyisobutyrate (pH 3.2–3.8)
FLOW RATE
PRESSURE
DETECTOR   Polarograph
SAMPLE   Rare Earths

Lu	1.3 h
Tm	1.6 h
Ho	2.2 h
Y	2.7 h
Tb	3.0 h

REFERENCE   J. F. Boissoneau, M. J. Repellin and A. Eglem (1980)
Analusis 8, 230.

(b) COLUMN   Corasil hydroxypropyl (90 cm × 2 mm i.d.)
COLUMN TEMP
MOBILE PHASE 10% ethyl acetate–cyclohexane
FLOW RATE   1.0 cm^3 min^{-1}
PRESSURE
DETECTOR   u.v.
SAMPLE

	k'
Al (acac)$_3$	6.17
Cr (acac)$_3$	4.70
Mn (acac)$_3$	0.90
Fe (acac)$_3$	0.91
Co (acac)$_3$	1.14
Ni (acac)$_3$	0.54
Cu (acac)$_2$	0.55
Cu (hepta-3,5-dionate)	0.19

REFERENCE   S. A. Matlin and J. S. Tinker (1979) *J. High Res. Chrom.,*
Chrom. Comm. 2, 507.

(c) COLUMN   Amberlite 200 (25–30 μm) (12 cm × 5 mm)
COLUMN TEMP 40°C
MOBILE PHASE 1st 0.06 mol dm^{-3} HCl (40–95%) acetone
                2nd 0.06 mol dm^{-3} HCl (57% acetone: 36% dimethyl-
                    formamide)
FLOW RATE   3.4 cm^3 min^{-1}
PRESSURE   16–25 bar
DETECTOR   Reaction detector
SAMPLE

Cd	2.5 min
Zn	4.6 min
Fe	6.5 min
Pb	10.0 min
Cu	13.0 min
U	15.0 min
Co	17.0 min

```
                    Fe    20.0 min
                    Mn    23.0 min
                    V     27.0 min
                    Ni    30.0 min
```

REFERENCE   G. Schwedt (1979) *Chromatographia* **12**, 613

(d) COLUMN   Lichrosorb RP-18 10m
     COLUMN TEMP 20°C
     MOBILE PHASE acetonitrile–water (65 : 35)
     FLOW RATE   1.5 cm^3 min^{-1}
     PRESSURE   45 bar
     DETECTOR   u.v. 254 nm
     SAMPLE   metaldiethyldithiocarbamates
                      Pb   6.2 min
                      Ni   7.7 min
                      Co   9.4 min
                      Cu   10.9 min
                      Hg   12.3 min

REFERENCE   J. Lehotay, O. Liska, E. Boardsteterova and G. Guchon
              (1979) *J. Chromatog.* **172**, 379.

(e) COLUMN   Lichrosorb RP-8 (25 cm × 4.6 mm i.d.)
     COLUMN TEMP
     MOBILE PHASE acetonitrile–water (70 : 30)
     FLOW RATE   1.5 cm^3 min^{-1}
     PRESSURE   38 bar
     DETECTOR   285 nm
     SAMPLE   Chromium VI diethyl dithiocarbamate   4.0 min

REFERENCE   G. Schwedt (1979) *Fresenius' Z. Anal. Chem.* **295**, 382

(f) COLUMN   Silica 13 μm (30 cm × 3.3 mm i.d.)
     COLUMN TEMP
     MOBILE PHASE heptane–dichloromethane–isopropanol (89.8 : 10.0 : 0.2)
     FLOW RATE   1.28 cm^3 min^{-1}
     PRESSURE
     DETECTOR   u.v. 254 nm
     SAMPLE   carborane π complexes

                                                       k'
                      1-C$_5$H$_5$-Co- 2,3- C$_2$B$_9$ H$_{11}$ 11.4
                      1-Et-C$_5$H$_4$-Co- 2,3- C$_2$B$_9$H$_{11}$ 4.4
                      1-Et-Co- 2,8-C$_2$B$_9$H$_{11}$ 0.5
                      guard column 60 mm × 3 mm i.d. Corasil II

REFERENCE   Z. Plzak, J. Plesek and B. Stibr (1979) *J. Chromatog.*
              168, 280

8.6 Miscellaneous

(a) COLUMN   Silica gel 7 + 10–25% 2-(2,4,5,7-tetranitro-9-fluorenylidene-aminoxy)propionic acid (0.2 m × 2.3 mm)

COLUMN TEMP

MOBILE PHASE   cyclohexane–dichloromethane

FLOW RATE   0.256 cm^3 s^{-1}

PRESSURE

DETECTOR   u.v.

SAMPLE

Pentahelicene	1.00
Hexahelicene	1.52
	1.69
Nonahelicene	3.20
	3.79

REFERENCE   F. Mikes, G. Noshart and E. Gil Av (1976) *J. Chem. Soc., Chem Comm.* 99.

(b) COLUMN   Zorbax SIL 6 μm (0.25 m × 2.0 mm)

COLUMN TEMP

MOBILE PHASE   gradient: 0.1% propan-2-ol in hexane to
1.0% propan-2-ol in dichloromethane

FLOW RATE

PRESSURE   1000 lbf in^{-2}

DETECTOR   u.v.

SAMPLE

Nitrobenzene	1.0 min
Methyl benzoate	1.4 min
α,α-Dimethylbenzyl alcohol	5.2 min
Cinnamyl alcohol	7.1 min

REFERENCE   R. C. Williams, D. R. Baker, J. P. Larmann and D. R. Hudson (1973) *Internat. Lab.* Nov. p. 39.

(c) COLUMN   Corasil I (0.5 m × 2.1 mm i.d.)

COLUMN TEMP

MOBILE PHASE   n-hexane

FLOW RATE   0.57 cm^3 min^{-1}

PRESSURE

DETECTOR

SAMPLE

C_2H_5HgCl	4.0 min
CH_3HgCl	9.5 min

REFERENCE   W. Funaska, T. Hanai and K. Fujimura (1975) *J. Chromatog. Sci.* 12, 517.

(d) COLUMN Porasil A (0.60 m x 2.0 mm i.d.)
COLUMN TEMP
MOBILE PHASE 0.5% pyridine—toluene
FLOW RATE 2.5 cm^3 min^{-1}
PRESSURE
DETECTOR atomic absorption
SAMPLE

Chromium hexafluoroacetylacetonate	2.5 min
Tris-(2'-hydroxyacetophenono)chromium	3.0 min
Chromium acetylacetonate	5.0 min

REFERENCE D. R. Jones and S. E. Manahan (1975) *Anal. Lett.* 8, 569.

(e) COLUMN Zipax + 1% tri-(2-cyanoethoxy)propane (3 m x 2.1 mm i.d.)
COLUMN TEMP
MOBILE PHASE hexane saturated with stationary phase
FLOW RATE 1 cm^3 min^{-1}
PRESSURE 600 lbf in^{-2}
DETECTOR u.v. 254 nm
SAMPLE Dinitrophenylhydrazones of:

n-Pentanal	8.5 min
n-Butanal	10.2 min
Acetone	10.2 min
Acetaldehyde	20.8 min
Formaldehyde	35.2 min

REFERENCE L. J. Papa and L. P. Turner (1972) *J. Chromatog. Sci.* **10**, 747.

(f) COLUMN Partisil 10 ODS 2 (25 cm × 4.6 mm i.d.)
COLUMN TEMP
MOBILE PHASE methanol–water–acetic acid (400 : 600 : 1)
FLOW RATE 2 cm^3 min^{-1}
PRESSURE
DETECTOR u.v. 273 nm
SAMPLE

o-Hydroxyphenylacetic acid	8 min
8-Hydroxycoumarin	14 min
7-Hydroxycoumarin	14 min
Coumarin	23 min
6-Hydroxycoumarin	11 min
5-Hydroxycoumarin	18 min
3-Hydroxycoumarin	28 min
o-Coumaric acid	21 min

Pre column of Co-Pell ODS (5 cm × 4.6 mm i.d.)

REFERENCE D. G. Walters, B. G. Lake and R. C. Cottrell (1980) *J. Chromatog.* **196**, 501

(g) COLUMN   RP-8 10 μm (25 cm × 4.6 mm i.d.)
   COLUMN TEMP
   MOBILE PHASE phosphate buffered tetrabutyl ammonium phosphate in
         methanol–water gradient 32% to 50% methanol at 3.6%
         min^{-1}
   FLOW RATE 1.5 cm^3 min^{-1}
   PRESSURE
   DETECTOR u.v. 254 nm
   SAMPLE   Phloroglucinol   2.8 min
         Phloroglucinol monosulphate   3.7 min
         1,2,3,5-Tetrahydroxybenzene-2,5-disulphate 8.5 min
         Phloroglucinol trisulphate   10.6 min
         2,3-Dibromo-4,5-dihydroxybenzyl alcohol 1′-4-disulphate
                       12.0 min

   REFERENCE M. A. Ragan and M. D. Mckinnon (1979) *J. Chromatog.*
         178, 505.

(h) COLUMN   Chiral phase based on fluoroalcohols attached to m
         Porasil (25 cm × 9 mm i.d.)
   COLUMN TEMP
   MOBILE PHASE 20% isopropanol in hexane
   FLOW RATE
   PRESSURE   u.v. 254 and 280 nm
   DETECTOR
   SAMPLE   Enantiomers of γ-2,4-dinitrophenyl 91 min
         γ-Butyrolactone   106 min

   REFERENCE W. H. Pirkle, D. W. House and J. M. Finn (1980)
         J. Chromatog. **192**, 143.

Subject Index

Compound Index